Quantum Physics

*A Beginners Guide to How Quantum
Physics Affects Everything around Us*

executed to present accurate, up to date, and reliable, complete information. No warranties of any kind are declared or implied. Readers acknowledge that the author is not engaging in the rendering of legal, financial, medical or professional advice. The content within this book has been derived from various sources. Please consult a licensed professional before attempting any techniques outlined in this book.

By reading this document, the reader agrees that under no circumstances is the author responsible for any losses, direct or indirect, which are incurred as a result of the use of information contained within this document, including, but not limited to, — errors, omissions, or inaccuracies.

Table Of Contents

MYTH #5: If you don't experience quick results, then the quantum jumps are not working

Chapter Thirteen: Summary/ Highlights of Quantum physics
Certain Particles Virtual in nature can appear and disappear randomly.
Superposition- The smaller a thing happens to be, the more likely it is to be present in different places all at once.
Quantum Tunneling- Micro Particles That Can Teleport
Entanglement Is When Tiny Things Can Communicate Faster Than Light

Conclusion

Introduction

I want to thank you for choosing this book, *'Quantum Physics - A Beginners Guide to How Quantum Physics Affects Everything Around Us'* and hope that you find this book informative and interesting in your quest to understand Quantum Physics and the role it plays in our everyday life.

Since you have chosen this book, it is certain that you have an interest in the field of quantum physics and look forward to understanding the concept better. For most of us, quantum physics is a complicated labyrinth that is meant for the geeks of the world; however, there is no denying the fact that the theories of quantum physics are extremely exciting.

Quantum physics as a term was not exactly a part of our daily parlance, but the Sitcom – The Big Bang Theory changed it all. When the characters in the series started to throw around big quantum theories and other scientific terms casually, a lot of us realized that quantum physics actually has a much stronger influence and role in our lives than what we actually know. I am not saying the show is the best place to start understanding quantum physics, but it can definitely be credited for making Quantum physics interesting to the layman, with no background in physics.

When we hear the term quantum physics, the first thought that comes to our mind is Einstein and his theory of relativity. Of course, it goes without saying that there is much more to quantum physics than that. Physics is an excellent medium of explaining a million different things starting from heating a cup of coffee to gravitational pull. There is no real limit in the discipline of physics. It involves matters that are as huge as the galaxy to things as small as neutrons. This book deals with the smallest side of it, which is the branch of quantum physics.

Throughout the course of this book, you will get a much better understanding of quantum physics starting from the basic concepts to some in-depth information. You will also see a lot of math and calculus in the book since quantum physics uses many concepts from those subjects. Don't dread reading through even though it might sound dreary and difficult. I don't intend to scare you with big equations and calculations, as this book will not make you a physicist. The sole aim of this book is to simplify quantum physics for the common man, who has no idea what it entails and how it affects our everyday life.

I have put the text together in a way that should make the subject matter much simpler to understand and maybe interesting to someone who normally hates science. I assure you that by the end you will have learned more than you normally do by just staring blankly ahead in a classroom. And if you are a curious student, you will definitely know more about quantum physics than before.

Quantum physics deals with the movement of microelements like electrons. Physicists have spent a lot of time and effort over the years in order to get this kind of in-depth knowledge of quantum physics. The branch of quantum physics is related to the smallest particles of matter or energy that exists. This is a more in-depth study than classical physics. As quantum physics has developed over the years, it has been able to explain the certain phenomenon that classical physics could not properly account for. Scientists took great interest in finding out more and correcting old norms, which were not essentially correct.

A lot of research has been and is currently being conducted in order to make quantum mechanics operational in our daily lives. For instance, if a proper quantum computer is made available to people, it will be able to solve many problems in a fraction of the time that it takes our standard computers. Such developments in the field of science are what will change the future. Quantum mechanics are also believed to be a crucial part in creating artificial intelligence. Complicated matters like national security will also benefit majorly from progress in quantum mechanics in cryptography. There are many such advantages that quantum physics can grant the world. You should learn more about it just like the scientists dedicating their lives to its growth.

Like I mentioned, this book is not meant to make you a physicist and is definitely not meant for you if you have a background in quantum physics. It intends to simplify the concept of quantum physics, which have been an enigma to the layman without a physics background. The things that you will learn in this book will definitely make you wiser in understanding how the world around us functions and the role quantum physics plays in making it work. One thing I can assure you is that by the end of the book, you will definitely feel smarter than before and will improve your general knowledge in things related to quantum physics.

Without further ado, let us take a deep dive into quantum physics and how it has a role to play in just about everything we do. I thank you once again for choosing this book and hope you find it an interesting read.

Chapter One: What is Quantum Physics?

Let us first understand what quantum physics really is. The word "quantum" is derived from Latin and means "how much". So, the quantum is in reference to the smallest particles of energy as well as matter that are studied in physics. Quantum physics is the branch that studies the conduct of the minutest levels of matter and energy. This is at the smallest microscopic levels including molecules, atoms and nucleons. It is important to separate this from the larger picture because it was proven that the laws for macroscopic objects are not the same for microscopic objects. Quantum physics also emphasizes on inter-particle interaction as well as electromagnetic radiation.

The development of the field of quantum physics can be attributed to people like Albert Einstein, Max Planck, Werner Heisenberg, and many others. Max Planck wrote one of the more prominent papers that contributed to its birth on blackbody radiation. This was as early as the start of the 1900's. Neils Bohr and Richard Feynman were also major contributors in this field. In fact, Johnston first used the term "quantum physics" in "Planck's Universe in Light of Modern Physics" in 1931.

Quantum physics is also called by other names such as quantum mechanics or even quantum field theory. This science has many subfields, but it is broadly known as quantum physics when it encompasses it all. It is also known as matrix mechanics or wave mechanical model. All this basically means the physics of the smallest atomic or subatomic levels of particle energy. Quantization, Wave-particle duality and the Uncertainty principle are very important aspects of quantum physics. The quantization principle explains how the momentum, energy and angular momentum as well as other properties of any bound system will be restricted to some discrete values. The wave-particle duality principles refer to how objects will have the characteristics of both particles as well as waves. The uncertainty principle states that there are certain limits to measuring quantities precisely.

Quantum physics is different from classical physics because the former deals with nature at a macroscopic level. Until quantum physics was developed, everything derived from classical physics would relate to the larger scale of matter. This branch of physics studies the elementary parts of matter. We learn the difference between the behavior of particles in an object and the behavior of the object. The quantum theory attempts to explain this unique behavior of matter and energy at a subatomic level. It also helps a great deal in understanding the behavior of particles that move at the speed of light or such extreme speed.

One of the key points in quantum physics is that it emphasizes that the observation of physical processes actually influence what takes place. The wave-particle duality tells us how light waves act as particles and particles can act like light waves. We learn about quantum tunneling through which matter goes from one spot to the other without travelling through the intervening space. Information can be passed in an instant through a vast distance. Quantum mechanics helps us realize that everything in the universe has many probabilities. It is just easier to deal with things on a larger scale than with smaller details.

Under quantum physics, various branches are singularly worked on. Consider the branch of Quantum Optics. In this, a scientist tends to focus on the behavior of light and photons. Unlike the concepts developed by Sir Isaac

Newton in classical physics, quantum physics shows that the behavior of each individual photon will have an effect on the overall light. In fact, this branch is what helped in the development of lasers as an application. Under Quantum Electrodynamics, we study the interaction of photons and electrons. Many scientists are trying to merge the concepts of quantum physics with the theory of general relativity. This is called the unified field theory. Some of these unified theories are Quantum Gravity, Grand Unified Theory, Loop Quantum Theory, String Theory, and the Theory of everything.

Another thing to keep in mind regarding quantum physics is that it is probabilistic. This is also the most controversial aspect of it. It is actually impossible to predict the outcome of any single experiment on a quantum system with absolute certainty. The outcomes of these experiments are predicted by keeping probabilities in mind for different outcomes. Repeated experiments are used to deduce the most probable outcomes. The "Born Rule" plays an important part in these cases but it is still being worked on to find something more absolute.

For the most part, quantum phenomenon is usually confined to the scale of atoms or fundamental particles. This is so that the mass and velocity is small enough for the wavelength to be big enough for direct observation. This is not so in case of large objects. Scientists are trying to make it possible for larger sizes as well.

All this has given you a starting point to understand what quantum physics is. As you read on, you will learn much more about how it originated and what it entails. Quantum mechanics have a huge scope for improving on a number of things more than classical physics.

The Three Revolutionary Principles of Quantum Physics

Over many decades, there was a lot that the math of classical mechanics could not explain. This was where quantum mechanics came into the picture. Quantum mechanics uses many mathematical concepts to explain such experiments. There is no single scientist who came up with quantum physics; many together contributed to its development and their theories only began getting acceptance around the early 1900's. There were three particularly revolutionary principles that laid its foundation in the scientific world.

Quantized Properties

Certain properties like color and speed are only supposed to occur in set spaces the same way a dial clicks on one number at a time. This challenges a fundamental assumption from classical physics that states that such properties should be on a smooth continuous spectrum. In order to explain that certain properties "clicked" a certain way, the term "quantized" was coined.

Wave-Particle Duality

Particles of light – Initially, the statement that light could sometimes behave like particles was harshly criticized. This was because for nearly 200 years, all experiments were based on and proved that light behaved in the form of a wave. Light could bounce off walls or bend around corners and have crests and troughs like those of an ocean wave. The source of light is like an object that is rhythmically dipped in the center of water. The color then emitted will correspond to the distance between crests that is determined by the ball's rhythmic speed.

Waves of matter – Matter was believed to only exist as particles. This was what decades of experiments had proven. It was then shown that matter could also behave like a wave.

The Uncertainty Principle

The uncertainty principle plays another major role in the development of quantum physics. In 1927, Heisenberg reasoned that there is a limit to how much precision of each property can be known. This is because since matter can act as waves, properties like the position and speed of an electron are complementary and there is a limit to determining the precision. According to him, the more precise the location, the less precise the speed of the electron is to be known. This also applies vice versa. It applies to everyday objects as well; the lack of precision is extremely tiny and is thus not noticeable.

These differentiate quantum physics from classical physics. So, let us look into it further in the text.

Chapter Two: History of Quantum Physics

Even as far back as 500 BC we see that Ancient Greek Philosophers speculated about the smallest part of a material or if matter can be divided at infinite. Around the late 1800's chemistry and the Brownian motion principle were used to answer such questions. Initially scientists defined the atom as the smallest particle of matter. Then there were studies on how atoms themselves were construed and what they were made of. It took more time to discover that atoms were made of protons and neutrons and that their own properties were different from that of the atom itself. And then later that quarks and gluons made them. The development of quantum physics was actually based on creating a viable model to study and explain the structure of atoms more minutely. When we study most theories, we see that classical theories tend to be insufficient or break down if we proceed towards a realm that is remote from normal experience. We see that classical Newtonian physics fail when we consider systems that travel too fast or it is about strong gravity, etc. It doesn't apply when instead of macroscopic objects we consider the microscopic world or even further in.

Around the 17th and 18th century, there was a lot of scientific inquiry about the wave nature of light. Scientists such as Robert Hooke, Leonhard Euler, and Christian Huygens had used experimental observations to propose a wave theory of light. The famous double-slit experiment performed by Thomas Young in 1803 also played a major role in people accepting this wave theory of light. Then Michael Faraday, in 1838, discovered cathode rays. After this Gustav Kirchkoff stated the blackbody radiation problem in 1859. In 1877 Ludwig Boltzmann suggested that a physical system can have discrete encrgy states and Max Planck gave his quantum hypothesis in 1900. One thing led to another and helped all this make sense altogether. The patterns observed in the observations of blackbody radiation were matching with what Planck hypothesized.

In 1896, its namesake Wilhelm Wien determined "Wien's law". The law determined the distribution of blackbody radiation. Using Maxwell's equations, Ludwig Boltzmann also reached this result independently. This underestimated radiance at low frequency and was valid just for high-frequencies. Planck who used Boltzmann's statistics and proposed "Planck's Law" later corrected the model. This Planck's law played a major contribution to quantum physics.

Some of the first few scientists who studied quantum phenomena were Pieter Zeeman, C.V. Raman, and Arthur Compton. Some prominent quantum effects are named after these brilliant minds. Robert Andrews Millikan experimentally studied the photoelectric effect, but Albert Einstein developed the theory. Similarly, Neils Bohr developed the theory of atomic structure though Ernest Rutherford's experiments discovered the actual structure of an atom. This theory was extended by Peter Debye to include elliptical orbits. This phase is referred to as the old quantum theory.

The Quantum theory was developed as different scientists made a series of discoveries or logical guesses. All these were sown together to create the theory of quantum physics. Some of the major steps towards it were:

- Max Planck's Black Body Theory in 1900

- Einstein's Light Quanta in 1905

- Bohr's Model of the Hydrogen Atom in 1913

- De Broglie's Hypothesis in 1924

These four contributions played a major role in advancing the field of physics in the world and, more particularly, the field of quantum physics. Let us look at some of the developments that led to the growth of this field separately from classical physics.

1. Thomas Young carried out the double-slit experiment to show the wave nature of light in 1801.

2. In 1896 Henri Becquerel discovered radioactivity.

3. In 1897, J.J. Thompson discovered the electron and its negative charge with his cathode ray tube experiments.

4. Ludwig Boltzmann was the founder of the Austrian Mathematical Society. In 1877 he suggested that the energy levels of a physical system like a molecule could be discrete. He founded this based on his studies in statistical mechanics and statistical thermodynamics and backed it with mathematical arguments.

5. Max Planck, a German physicist, introduced the idea of energy being

"quantized", in 1900. This was reluctantly done in his attempt to derive Planck's Law, as it is still known. This formula was created for the frequency dependence of energy that is emitted by black bodies. The Wien approximation from earlier can be derived from Planck's law as well.

6. Planck's quantum theory helped Stefan Procopius and Neils Bohr in calculating the magnetic moment of electrons. This was later called the magneton. It was also possible to make similar computations of the magnetic moments of protons and neutrons.

7. Einstein then explained the photoelectric effect in 1905 by postulating that all electromagnetic radiation including light could be divided into a number in energy quanta, which were localized points in space.

8. In 1909, Robert Millikan showed that electric charge occurs as quanta with his oil drop experiment.

9. In 1911 the planetary model of the atom was developed when Ernest Rutherford showed that the plum pudding model of

atom was invalid.

10. The Bohr used quantization to explain the spectral lines of hydrogen atoms in 1913.

11. In 1912, Henri Poincare thoroughly discussed Planck's theory in this paper "Sur la theorie des quanta".

12. All these together are referred to as the old quantum theory.

13. Louis De Broglie was a French Physicist who put forward another theory in 1923. He suggested that particles could have wave characteristics and waves could have particle characteristics. He developed this theory from that of special relativity and applied it to a single particle.

14. Modern quantum physics was born when physicists Werner Heisenberg, Pascual Jordan and Max Born used De Broglie's approach to develop matrix mechanic in 1925.

15. Simultaneously, Erwin Schrodinger invented the wave equation and Schrodinger equation. He also put forward the thinking experiment that everyone knows as Schrodinger's cat.

16. In 1927, Heisenberg put forward the uncertainty principle.

17. The Copenhagen interpretation was also taking shape around the same time. You will read more about this further in the book.

18. Paul Dirac started working on unifying special relativity with quantum mechanics. He proposed the Dirac equation for electrons. This equation did one up on Schrodinger equation by being able to achieve the relativistic description of an electron's wave function. The Dirac equation also predicts electron spin and he predicted the existence of Positrons. Dirac also established the use of the operator theory.

19. John Von Neumann was a Hungarian Polymath who formulated an arduous mathematical base for quantum physics. In his book published in 1932, he published the theory of linear operators on Hilbert spaces. This work is still valid.

20. Quantum field theories started developing when scientists began applying the quantum mechanics to fields and not just

singular particles. P.A.M. Dirac and P. Jordan are some of the contributors in this part. This research helped the formulation for quantum electrodynamics in the 1940s by F. Dyson, R.P. Feynman, S. Tomonaga, and J. Schwinger. Quantum electrodynamics served as a model for quantum field theories later on. It described a quantum theory involving electrons, positrons and electromagnetic fields.

21. In 1955, Clyde L. Cowan and Frederick Reines verified the existence of neutrinos.

22. In the 1960s the theory for quantum chromodynamics started being formulated. The present-day theory was refined in 1975 by Gross, Politzer, and Wilczek.

23. Glashow, Salam, and Weinberg received the Nobel Prize in 1979 for physics when they showed a single electroweak force would be obtained if a weak nuclear force were merged with quantum electrodynamics. This work was built on the pioneering research of Higgs, Schwinger, and Goldstone.

As you can see, science is a field that continually redefines things that we were previously sure or unsure of. The field of quantum physics was established based on this. The works of decades by many scientists has given us the knowledge of the atomic and subatomic world that we now have. This field is still in a stage of infancy and has a long way to go before we know more in detail and with perfect accuracy.

Chapter Three: Theories of Matter

Matter was known to be in two forms at the end of the nineteenth century; one was particle and the other form was waves.

Particles were like a localized mass that flew around like tiny bullets. Out of all fundamental particles, electrons were the most investigated. In 1896, Thompson conducted his cathode ray tube experiment. He found that the cathode rays in cathode ray tubes were deflected by magnetic and electric fields, like tiny bits of matter charged with electricity. Atoms were also particulate in matter.

The other form was matter in waveform. The most properly investigated of these was light or electromagnetic waves. Many scientists in the seventeenth century gave accounts of how light consisted of a shower of tiny little corpuscles. Newton's corpuscular view remained dominant at the time even though some had suggested the account of wave-like behavior. This only changed at the beginning of the nineteenth century when those like Thomas Young drew different inferences from their studies.

The two-slit experiment is the most recognized

interference effect. In this experiment, waves of light strike a barrier that has two holes in it. The waves are shown as parallel wave fronts that move up the screen. From the two slits, secondary waves will radiate out and interfere with one another. This forms the cross-hatching pattern of interference that is characteristic. The same patterns can be observed on the surface of a pond when pebbles are dropped in the water and ripples are caused. These ripples will have the same pattern observed here. The manner in which the waves combine is of importance in this kind of interference experiment. It is because waves may add up to two ways that these patterns will arise.

The phases of waves in constructive interference will add to form combined waves of greater amplitude. Every part of each wave will line up to interfere constructively in every place. Destructive interference causes the phases to occur in a way that they subtract to cancel out. Every part of the waves will line up in a manner that they interfere destructively in every part.

Ordinary cases of interference like the two-slit experiment exhibits both constructive and destructive interference in different parts where intersection of the waves happen. Complicated interference patterns will be observed in such cases. If you think of waves as a kind of displacement in mediums, they can be easily understood. For instance, in the ocean a wave will have peaks where the water is above sea level and troughs where the water is below sea level. When two waves meet and if their peaks coincide, a peak with combined height will result from these. This can be understood as constructive interference. If the peak of a wave coincides with the trough of another, the two will cancel each other out. This pattern is called destructive interference.

The explanation of interference was very compelling for Maxwell. He considered it good evidence in case of ether. According to him, light had to be a displacement in something in order for it to have peaks and troughs that can be cancelled out. That carrier could be ether. He surmised that if light was made up of many tiny corpuscles, then it might be impossible to combine two and cause self-annihilation. The ether theory died out but there was a possibility for something better. Somehow, light came in a form that could cancel out other light waves. This was an early indication for the later deductions of quantum theory.

All of this was at the end of the nineteenth century and formed the foundation for more.

The quantum theory of matter is a microscopic explanation of the properties of solids and liquids based on quantum mechanics. Properties like magnetism and superconductivity actually have no explanation without the quantum theory. Some other properties only have phenomenological descriptions without the help of this theory. Using the quantum theory, you can comprehend the approach that is really required in order to understand everything.

Quantum physics of matter is different from traditional theory because the solutions of appropriate quantum mechanical equations are not known despite knowing the constituents as well as interactions that take place. The problem is because the complexity of the interactions between various components causes the formation of new complex structures like crystals, glass, superconductors, magnets, etc. This means that new conceptual theories have to be brought up to create predictive theories of matter. The normal technique for approaching this problem for condensed matter systems is by trying to reduce the number of variables to a manageable number that can still describe the essential physics.

The quantum theory along with the theory of relativity together forms the theoretical basis of modern-day physics. The theory of relativity assumes importance when considering very high speeds involved. Similarly, the quantum theory is needed when you consider very small quantities in the scale of atoms, molecules, and elementary particles. Many aspects of quantum theory have sparked a lot of debate especially due to those like the uncertainty principle. The theory of relativity was primarily the work of Albert Einstein. The quantum theory has been developed over decades from the contributions of many scientists.

Chapter Four: Superposition, Interference, Decoherence

In an ocean, big waves can have tiny ripples overlaid on them. In the same way, superimposition can exist in the subatomic world. The existence of two or more waves is permitted by Schrodinger's theory. If light shines on a transparent glass, a wave could correspond with a photon that passes through the glass and another could correspond with a photon that bounces back. On the other hand, the waves could also be superposed so the photon can be transmitted and reflected at the same time. This means that the photon can be on both sides of the glass at the same time. This shows the possibility of the potential of infinite number of waves that are superposed. This means that microscopic particles could actually exist in a number of places at the same point in time and have numerous ways to potentially behave.

Interference is the ability of two waves passing through one another to mingle and thus reinforce each other at the coinciding crests and cancel each other where the crest coincides with a trough. It is similar to the interference caused by ripples in water. Thomas Young's double slit experiment has a modern incarnation where a feeble light source gives out one photon at a time. This gives the same interference evidence even when no waves interfere. This can happen only if each photon goes through the splits simultaneously and interferes with itself. This can only be possible if the photon is in a state of superposition. This experiment can be performed with any subatomic particles and demonstrates that every particle has both wave and particle aspects. Quantum computers can combine multiple calculations into one by using interference phenomenon.

In reality, superposition cannot actually be observed, the consequence of it after interference is the only thing that is visible. So, we can't observe an atom in indeterminate state or superposition. Only when measurement takes place can the physical reality be determined and the situation be solidified.

Decoherence is the problem in observation and measurement of superpositions. Attempting to measure or obtain information from quantum superpositions leads to decoherence. This effectively destroys the superposition and it is reduced to a single state or location. The individual states also lose their ability to hinder with each other's role. Decoherence causes the downfall of a quantum wave and makes the particle settle into the state that is observed by classical physics. Thus, there is a transition of behavior from quantum to classical. It is one of the main reasons that quantum theory applies only to the subatomic world in practice. Schrodinger's cat experiment illustrates the problem caused by decoherence. You will read more about it later.

It would be impossible to isolate objects in the macroscopic world from interacting with the environment. In such a case, there are more than trillions of photons to consider. The interaction of the quantum objects with the environment gives us the classical objects that we easily comprehend. We see that in practice, we do not observe quantum systems directly, but we see the effect that the quantum systems have on their environment.

Chapter Five: Quantization

Classical physics explained many phenomena using theories that at that time seemed to be quite accurate. Over the years, these classical theories fell apart and could not accurately account for their explanations. These were then proven using the dynamics of quantum physics. It began with the failure of explaining the blackbody radiation. You should know that every tangible body emits electromagnetic radiation or light energy. The amount of this radiation depends on various factors such as the temperature of the body and its color. The higher the temperature of the body, the lighter it emits. The reason we cannot see the light emitted from bodies at room temperature is that this light will be from the infrared spectrum which is invisible to the naked eye.

Let us understand what quantization really is. In physics, it refers to the process of transition from classical view to a newer understanding that is known as quantum mechanics. It involves the construction of the quantum field theory from classical physics. Field quantization is when we refer to the quantization of an electromagnetic field. Here, photons are called quanta.

Quantization involves conversion of classical fields into operators that act in a quantum state of field theory. Vacuum state is the lowest known energy state. A theory is quantized in order to infer the properties of any material or particle using quantum amplitude computations. These can be very complicated and need to be dealt with subtly, known as renormalization. If ignored, this can cause deductions of useless results. Quantization procedures require renormalization methods. Canonical quantization was one of the first such methods.

Canonical quantization of field theories is equivalent to construction of quantum physics from classical physics. The classical field is the canonical coordinate here and is a dynamical variable. The canonical momentum is the time derivative. A commutation relation is introduced between these, which is the same as that between position and momentum in quantum mechanics. The field is converted to an operator using combinations of annihilation and creation operators. The field operator acts upon the quantum states of theory. Vacuum state is the lowest energy state. This procedure is also known as the second quantization. It can be applied for quantization of any field theory from fermions to bosons. Any field symmetry is also fine. It is not good enough for some quantum field theories like quantum chromodynamics since the vacuum state is portrayed too simply. Some other types of quantization are Geometric quantization, Deformation quantization, covariant canonical quantization, path integral quantization, etc.

The black body was a simplified model created to find the spectral composition that was emitted by a body with respect to its temperature. A hypothetical body absorbs any light that strikes it and when it reaches an equilibrium temperature, it re-emits that light at the same rate. The composition of this spectrum could not sufficiently be explained using classical physics. According to the classical theories, a black body would thus emit huge amounts of energy in high-frequency radiation and release an infinite quantity of energy. This contradicts principles like the conservation of energy and many more. They knew that a new model was required to explain it.

This is where Max Planck came in. He was a German Physicist who came up with the idea that bodies emit light through small quanta packets and not continuously as was thought before.

[E=h.f where h=6.626 x 10-34, E=photon energy, f=photon frequency]

The above equation gives the size of the small packets called quanta and states that the energy of a photon is directly proportional to its frequency where there is a constant. This constant is now known as Planck's constant. Planck could not give a proper justification for this quantization, Einstein's explanation of the photoelectric effect in later years allowed scientists to take Planck more seriously.

Chapter Six: Wave-Particle Duality

The wave-particle duality concept of quantum physics states that any matter may not just be described as particles but also as waves. This principle puts on emphasis in showing how classical concepts cannot fully comprehend or explain the behavior of objects on a quantum scale. The work of scientists like Albert Einstein, Max Planck, and Neils Bohr discovered how particles describe a wave-like nature and also vice versa. This applies for both elementary and compound particles. Wave properties are hard to detect in macroscopic objects due to their short wavelengths.

According to Neils Bohr the duality paradox is a fundamental fact in nature. Any quantum object may sometimes behave like a particle and sometimes like a wave depending on the physical setting. It is an aspect of the complementarity concept. Werner Heisenberg was another scientist who took this further.

Particles of Light

Till the late 1800's Maxwell's theory of electromagnetic radiation was considered a basic fact and there was no questioning it. Then in 1900 Max Planck introduced the concept of light carrying energy in small quantities. At that time, this was a very foreign and unsettling announcement since it made scientists unsure of something that they did not even question. Planck proposed that those quantities are made up of energy increment hf where h is Planck's constant and f is the radiation frequency. Slowly as scientists like Albert Einstein began working with this, the quantum theory of light slowly emerged and took over.

Waves of Matter

In 1924, a French Physicist with the name Louis De Broglie, used Einstein's equations for the theory of special relativity and showed that particles could also exhibit wave-like characteristics and vice versa. Two other scientists further utilized this reasoning in 1925 that tried to explain how electrons moved in an atom. At that time this could not be explained using classical physics. The wave-particle duality of light had made scientists start questioning if matter also could only act as particles. German physicist Werner Heisenberg then explained this by developing matrix mechanics. Erwin Schrodinger was another physicist who then developed wave mechanics and later showed that both of these were equivalent. This led to the Heisenberg-Schrodinger model of an atom where electrons act like waves around the atom's nucleus. According to this model, electrons obey a wave function and rather than orbits, they occupy orbitals. This is unlike the Rutherford-Bohr model of an atom where orbits can only be circular. Later in 1927, wave mechanics were developed to show that atomic orbitals could combine and form molecular orbitals.

First there were theories and then there were experiments, which resulted in observations that led to making the theories of quantum physics more substantial. For instance, Max Planck found a solution to the blackbody radiation problem in the 1900's, which classical physics could not explain. Quantum physics has been very important over the years. It has found application in quantum chemistry, quantum computing, electron microscopy, resonance imaging, quantum optics, medical and research imaging, etc.

History of Viewpoints on Wave and Particle

According to Democritus, light and everything else in the universe is composed of subcomponents that are indivisible. Alhazen was an Arab scientist in the 11th century who comprehensively wrote a treatise on optics that described refraction, reflection, as well as the operation of a pinhole lens. Alhazen asserted on the point that rays of light were composed of particles. Rene Descartes, in 1630, spoke on the opposing wave description; he showed that wave-like disturbances could be demonstrated to recreate the behavior of light. Isaac Newton later developed his corpuscular hypothesis where he argued that the straight lines of reflection show that light is made up of particles. He explained that only particles could travel in such straight lines and the refraction occurred when these particles got accelerated when entering any denser medium. Scientists like Robert Hooke and Augustin Jean Fresnel refined this viewpoint on wave using mathematics. They showed that light travelled at different speeds in different mediums, and this explained the phenomena of refraction according to them. The Huygens-

Fresnel principle was also supported by the discovery of the double slit interference that Thomas Young discovered in 1801. The wave view took time to displace this particle view in around the 19th century when scientists took it more seriously. This was because the wave nature could explain polarization but not the other views.

James Clerk Maxwell's discovery showed that any type of light was an electromagnetic wave of different frequencies. The wave theory was very successful in the 19th century but at the same time the atomic theory of matter was also prevailing. Much advancement in chemistry was made such as the discovery of new elements or compounds. Antoine Lavoisier deduced the law of conservation of mass as well. By discovering diatomic gases, Amedeo Avogrado finished the basic atomic theory. This allowed the correct deduction of the molecular formulae of most compounds known at the time. Then Mendeleev created his periodic table when he saw the order of recurrence in properties.

At the end of the 19th century, there was more study of what the atom itself was. Electricity was first considered fluid, now it was shown to be comprised of electrons. In 1897, J.J. Thomson demonstrated this using a cathode ray tube where he showed that an electric charge could travel through a vacuum. Since vacuums are vacant spaces with no medium, the only explanation was that a particle was carrying a charge and travelling through the vacuum. Thus, the electron was now a subject of big discussion since it went against the decades of studies where electricity was treated as fluid.

Max Planck then published an article where he demonstrated the successful reproduction of the spectrum of light from a glowing object. Later Einstein said that electromagnetic radiation was not the energy of radiating atoms but is quantized itself. Blackbody radiation was a concept that could not be explained through classical theorems. When an object's heat causes electromagnetic radiation to be emitted it is called blackbody radiation. According to the equipartition theorem, any object's energy would be equally partitioned amongst its vibrational modes. The same reasoning could not be applied to such an object's electromagnetic emission. It was a long-known fact that thermal objects tend to emit light. Scientists wanted to explain this phenomenon on the basis of the fact that light was made of waves of electromagnetism.

This was the blackbody problem, as it became widely known. There was a problem when it was realized that if an equal partition of energy were received by each mode, all the energy would be consumed by the short wavelength modes. This was clear when Rayleigh-Jeans law predicted the intensity of emissions of long wavelength and also infinite total energy since the intensity of short wavelength diverges to infinity. This was referred to as the ultraviolet catastrophe.

Max Planck also hypothesized that a blackbody's light frequency depends on the frequency of the oscillator emitting it. He also said that the energy of such oscillators increased with the frequency in a linear manner. This made sense since macroscopic oscillators have similar functioning. Planck avoided arguments by giving high-frequency oscillators equal partition and thus producing fewer oscillators as well as less light emission. In Maxwell-Boltzmann's distribution suppressed the oscillators of low frequency and energy by thermal jiggling from high-energy oscillators and this increased their energy as well as frequency. The revolutionary aspect of Planck's black body was that it inherently depended on integer oscillators in thermal equilibrium with electromagnetic field. A quantum of light was created when the electromagnetic field received the entire energy from these oscillators. While Planck's theory of blackbody radiation was intentional, he unintentionally also created an atomic theory of light. He had denounced the particles of light as a limitation of this study.

Philipp Lenard had discovered that the energy of ejected electrons depended on their frequency and not the intensity of the light. Low light falling on a metal object would cause the ejection of electrons with low energy. In case of a more intense beam of the same low frequency, the result would just have more electrons ejected but of the same low energy. This theory was at odds with the theory of there being a continuous energy transfer between matter and radiation. Also, Albert Einstein used Max Planck's black body example to produce a solution for the photoelectric effect problem.

The photoelectric effect could be explained quite simply if Planck's energy quanta is used and it is considered that at a given frequency, electromagnetic radiation could only transfer energy to matter in integer multiples of hf energy quantum. The low energy electrons are ejected from low frequency light due to absorption of single photons. Thus, the energy is not increased even if the intensity of light is increased at the same low energy level. Electrons of higher energy will only be ejected if the frequency of light is also increased since this will also increase the energy of the photons. If you use Planck's constant h, the energy of ejected electrons will increase linearly with the increase in frequency when you determine the photon's energy. In this case, Planck's constant will be the gradient of the line. All this was not confirmed till 1915 when experimental results were produced in perfect accord with the predictions of Einstein. Robert Andrews Millikan did this; he also determined the electron charge. Until the photon antibunching effect was discovered, the existence of photons was not proven. In fact, Einstein's 1921 Nobel Prize was for his basic yet revolutionary work on quantized light and not for his other more complex work. The light quanta he spoke of was not labeled as photons till 1925, it still represented the wave-particle duality concept.

Photoelectric Effect

Einstein's photoelectric effect caused a lot of

trouble when he stated it in 1905. At that time, he suggested the existence of quanta of light or photons that have particulate energy.

Einstein's statement was as follows:

"According to the assumption to be contemplated here, when a light ray is spreading from a point, the energy is not distributed continuously over ever-increasing spaces, but consists of a finite number of 'energy quanta' that are localized in points in space, move without dividing, and can be absorbed or generated only as a whole."

This was considered one of the most scientifically revolutionary statements of the twentieth century. The energy quanta that he mentioned were later to be labeled by Gilbert N. Lewis as "photons" as they are still termed. As per Einstein's idea each photon had energy in quanta. It was a remarkable theory and was the effective solution for the problem of blackbody radiation attaining infinite energy.

It was observed that an electric current would be produced in a circuit on shining light on certain metals. He assumed that this light was moving the electrons out from the metal and this was causing the flow in current. When he used an element like Potassium, a dim blue light caused a current but not the brightest red light. The classical theory stated that a light wave's strength or amplitude was proportional to the brightness, hence the bright red light should have been able to produce a large current. This was not the case. So, Einstein explained it by stating E=hf. Here E is the amount of energy that is related to f the frequency of light with h as Planck's constant. Only photons with energy above a particular threshold could knock electrons free. No amount of light below this threshold could release electrons. Violating this law would only be possible by using lasers of high intensity, which were not even invented at the time.

Wave Behavior of Large Objects

Just like demonstrations on photons and electrons for wave-like properties, experiments on neutrons and protons have also been conducted. Estermann and Otto Stern's experiment in 1929 is one of the more prominent ones. Other people who conducted such experiments say that larger particles behave like waves just like smaller ones.

In the 1970's the neutron interferometer was used to carry out experiments showing the gravity's action in relation to wave-particle duality. A lot of the mass of a nucleus or any matter is provided by neutrons. Neutrons act as quantum mechanical waves in a neutron interferometer where they are subjected directly to the force of gravity. It was the first time the self-interference of a quantum wave of huge fermion was confirmed experimentally.

Researchers from the University of Vienna reported the diffraction of C60 fullerenes in 1999. These fullerenes were a comparatively much larger object since their atomic mass was around 720 Dalton. The incident beam had De Broglie wavelength of 2.5 pm and 1 nm diameter of the molecule was hundreds of times larger. The same group also demonstrated the wave nature of tetraphenylporphyrin in 2003. By 2011, the Kapitza-Dirac-Talbot-Lau interferometer could be used to demonstrate interference of heavy metal molecules of 6910 Dalton. It is unclear if objects with weight more than Planck mass have De Broglie wavelength and it is also experimentally unreachable.

Planck's Law

Planck's law is a very important part of physics. This law is also known as the basis of quantum theory. Planck's law states that the energy of electromagnetic radiation is composed of units called quanta or photons and each of these has energy equal to h.f, where h is Planck's constant and f is the radiation frequency. The law essentially defines the density of any electromagnetic radiation that is released by a black body at a given temperature T in thermal equilibrium. The law got its name from Max Planck since he was the one who proposed it. This law has been a pioneering discovery in quantum physics.

Planck assumed that radiation sources are made up of atoms that are oscillating, and this vibrational energy of each oscillator could have any series of discrete values but not any value between them.

Formulation

Here, the symbols stand for the following:

E_λ = the energy radiated by a cavity of blackbody per unit volume in wavelength interval λ to λ + $\Delta\lambda$

H= Planck's constant

C= the speed of light

K= the Boltzmann constant

T= the absolute temperature

Planck's constant is also a fundamental constant in physics and participates in a massive number of formulations that are related to quantum physics. Written as "h", the constant was introduced by Max Planck in 1900 when he stated Planck's law.

E=hv

Here E is the energy of each quantum, h is the constant and v is the frequency of radiation. The energy of a photon equals Planck's constant times the radiation frequency. A modified form of Planck's constant is called h-bar, where h is divided by 2π. This h-bar is the quantization of angular momentum; the value of angular momentum can only be a multiple of it.

The value of Planck's constant is derived from

time multiplied by energy. Numerically it is given as 6.62607004 x 10^{-34} joule second and there is a standard uncertainty value 0.000000081 × 10^{-34} joule second. Planck's constant is usually called the elementary quantum of action.

Planck's radiation is considered thermal radiation due to its dependence on temperature. It has maximum intensity at a specific wavelength depending on temperature. At high-temperatures, the infrared radiation increases and comes out as heat while the body will glow red. Temperatures even higher than this will cause the body to glow yellow or bluish white and emit short wavelength radiation in significant amounts. The emission of the sun is peaked in the visible spectrum since it emits both infrared as well as ultraviolet radiation. Regardless of chemical composition and surface structure, Planck radiation is the maximum radiation emitted by any body in thermal equilibrium. Passage of radiation between media across an interface is characterized by emissivity of the interface. This is symbolized with ε. Generally; it will be depending on physical structure, chemical composition, temperature, wavelength, angle of passage, as well as polarization. The emissivity of any natural surface will be between ε = 0. A blackbody is any that interfaces with a medium which absorbs all radiation that is incident upon it and has ε = 1. A small hole in the wall of an enclosure can be used to represent a blackbody. This is maintained at uniform temperature and has opaque walls that are not perfectly reflective at every wavelength. Inside this enclosure, the radiation follows Planck's law at equilibrium and thus the radiation coming out of that hole will also follow Planck's law.

In case of a photon gas, the temperature determines the pressure, spectral radiance, and energy density. In a material gas, the mass and number of particles play a significant role. The second law of thermodynamics will cause photon energy distribution to change if there are interactions and approach Planck distribution. The photon gas must be Planckian for this to not occur. In this kind of approach, the photons will be created and destroyed in right numbers with right energies to fill a cavity with Planck distribution until the equilibrium temperature is reached.

Spectral radiance is represented as $B_v(v, T$ as a function of frequency and temperature. In the SI system the units are $\underline{W} \cdot \underline{m}^{-2} \cdot \underline{sr}^{-1} \cdot \underline{Hz}^{-1}$. The

Blackbody is called a Lambertian radiator when Planckian's radiation spectral has the same value for any course or angle of division.

Via the Equipartition theorem, classical physics led to the Ultraviolet catastrophe that predicted that total blackbody radiation intensity is infinite. If it is assumed that radiation is finite, then some aspects of Planck distribution like the Stefan Boltzmann law and Wien displacement law can be accounted for by classical thermodynamics. If matter is present, then quantum mechanics will provide a good account. If matter is absent, then quantum field theory is needed since non-relativistic quantum physics with fixed article numbers do not provide sufficient accountability.

According to Planck's law, radiation is viewed as a gas of massless, uncharged, bosonic particles in thermodynamic equilibrium, i.e. photons. These photons are seen as carriers of electromagnetic interaction amongst electrically charged elementary particles. Photons are created and destroyed in right numbers and with right energies. Their number is not conserved. In thermodynamic equilibrium the internal energy density of a photon gas is completely determined by temperature. The pressure on the other hand is entirely dependent on the internal energy density. In thermodynamic equilibrium for material gases, internal energy is dependent not only on temperature but also the respective number of molecules and specific characteristics of those different molecules.

At a given temperature for material gases, the internal energy density as well as the pressure can vary because different molecules can independently carry different excitation energies. Planck's law actually arises as a limit of the Bose-Einstein distribution. For massless bosons like photons, there is zero chemical potential and the Bose-Einstein distribution is reduced to the Planck distribution. The Fermi-Dirac distribution is another fundamental equilibrium energy distribution. It describes fermions like electrons in thermal equilibrium. Both the distributions are different because multiple fermions can occupy the same quantum state but multiple bosons cannot. The difference becomes irrelevant when the density is low because the number of quantum states available is large per particle. The Bose-Einstein and Fermi-Dirac distribution reduce to Maxwell Boltzmann distribution in low-density limit.

Spectral Dependence on Thermal Radiation

Radiative heat transfer and conductive heat transfer are different from each other. Radiative heat transfer only filters through a definite band of radiative frequencies. When a body gets hotter, more heat is radiated from it at each frequency. In thermodynamic equilibrium, there is only one temperature in a blackbody and it must be shared by radiation of any frequency. Balfour Stewart also discovered that a lamp black surface emits the maximum amount of radiation compared to any other surface. Kirchhoff further deduced that spectral radiance of such cavities in thermodynamic equilibrium has to be a unique universal function of temperature. He suggested an ideal blackbody, which would be interfaced with the surroundings in a way that all radiation that fell on it would be absorbed. The thermal radiation emitted from such a body at thermodynamic equilibrium would have the unique universal spectral radiance as a function of temperature. Kirchhoff's law of thermal radiation used this as insight.

De Broglie Wavelength

Louis De Broglie was a French Physicist who made a bold assertion in the 1923 doctoral dissertation. De Broglie proposed that if you consider the relationship of wavelength lambda and momentum p given by Einstein, it would the wavelength of any matter:

Lambda=h/p where h is Planck's constant.

This wavelength is known as De Broglie's wavelength. This wavelength is manifested according to wave-particle duality in all particles in quantum physics and determines the probability density of locating an object at any given point of the configuration space. De Broglie's wavelength is inversely proportional to momentum of the particle. Instead of the energy equation he decided to choose the momentum equation for a particular reason. He said that it was because it was unclear if E should represent kinetic, total, or total relativistic energy. The energies will be the same if it was photons but it was not the same for matter.

De Broglie assumed that the same relations are valid for particles as they are for photons. Unlike photons that always move at the same velocity of the speed of light, according to special relativity, the momentum of particles depend on the velocity and mass.

The wavelength hypothesis also got experimental confirmation. Clinton Davisson and Lester Germer were scientists of Bell Labs. In 1927 they performed an experiment wherein electrons were fired at a crystalline nickel target. The predictions made by De Broglie wavelength were matched by the diffraction pattern that resulted from this experiment. His theory earned De Broglie a Nobel Prize in 1929. The scientists who experimentally proved it and discovered electro diffraction also received the Nobel Prize in 1937 for their work. There were also other experiments that hold De Broglie's wavelength hypothesis as true. This includes the double slit experiment. The wavelength is also true for large molecules to the size of 60 carbon atoms or more.

The main significance of the hypothesis was that it showed wave-particle duality to be a fundamental principle of radiation as well as matter and not just an aberration of light. If you correctly apply De Broglie wavelength then it is also possible to describe material behavior using wave equations. The De Broglie wavelength is as such a crucial part of particle physics.

There are realistic limits for using De Broglie's hypothesis. In case of macroscopic objects, the wave aspect would be so tiny that it would not be observable to any real use. It can be done but just for those who want to particularly know of it but not in any useful sense.

Chapter Seven: Uncertainty Principle

The third important principle is Heisenberg's uncertainty principle. It is also called the indeterminacy principle and was postulated by Werner Heisenberg in 1927. This uncertainty principle states that even in theory, it is not possible to accurately measure the position or velocity of an object at the same time. He postulated that the concepts of exact position and velocity have no meaning in nature. Normal experience would not let you deduce this principle. For example, it is quite simple to calculate the position and velocity of an automobile. This is because the uncertainties that this principle implies are actually too small to even be observed.

The product of uncertainties, thus, becomes significant only for extremely small amounts of atoms or subatomic particles. According to Heisenberg's principle, the product of velocity and the uncertainties in position is either equal to or greater than $(h/(4\pi)$. Here h is Planck's constant and has the value 6.6×10^{-34} joule second. Attempting to be precise in measuring the velocity of subatomic particles like electrons will actually knock it around in a way that a simultaneous measurement would be without value. This would not occur due to inaccurate instruments or by the observer's fault, but because of the naturally intimate connection that particles and waves of the subatomic dimensions have.

This simple uncertainty principle actually explains why atoms do not implode and how the sun manages to shine. It tells us how there is a fundamental limit to how much we can know about the nature of particles and such details of nature. This principle adds a layer of uncertainty to quantum theory as a whole. Uncertainty is not something to worry about and it does not take away from the value of quantum physics but only adds to it.

One of the more prominent implications of this principle is that it says that a vacuum is not the absence of everything. According to quantum theory, there are particles in a vacuum as well. It is uncertain how much matter can exist in a space and at any uncertain time particles can appear in a vacuum as well. This can be explained by the principle because it can be considered that for extremely short amounts of time, the energy of a quantum system can be uncertain and at this point the particles can come out from the vacuum. These particles might appear for a very short time and then destroy each other and be within the laws of quantum physics.

Each particle is associated with a wave and also exhibits wave-like behavior. The particle has higher probability of being found where the wave's undulations are greatest. The wavelength becomes more undefined if the undulation of that wave becomes more intense and this affects the momentum of that particle. A particle wave with properly defined wavelength will be spread out and the associated particle would have precise velocity but could be anywhere. When the wavelength is indeterminate, the associated particle will have a fixed position but uncertain velocity. Therefore, we see that a large uncertainty exists in measuring one observation when the other is more accurately measured.

Heisenberg's principle is expressed either in terms of momentum or position of a particle. The product of mass into velocity will give the momentum of the particle. So, you can deduce that the product of position into the uncertainties in momentum will equal to either $h/(4\pi)$ or even more. This same principle also applies to other conjugate pairs such as time and energy.

The uncertainty principle actually helps to explain many things that classical physics cannot account for. For instance, let us consider atoms. In atoms, a positively charged nucleus is surrounded by negatively charged electrons. It would be normal to expect that the opposite charges would attract each other and lead to the collapse of everything. With the help of the uncertainty principle, you can understand that the electrons position can be precisely known if it is very close to the nucleus since the error in measuring this would be very minute. There would be a huge error while measuring its momentum. This would mean that the high speed of the electron could lead to it flying right out of the atom.

Alpha decay is also explained with the help of Heisenberg's principle. It is a type of nuclear radiation. When two protons and neutrons are emitted by a heavy nucleus, they are called alpha particles. A lot of energy would be required to release these from the bond that keeps them inside the heavy nucleus. The position of the alpha particle is not so well defined since it has a very defined velocity. There is a very minute chance of the particle finding its way out of the nucleus at some point, but it is close to zero. In case this happens, the alpha particle could escape, and this would demonstrate radioactivity. This process is known as quantum tunneling since the particle couldn't leap over the energy barrier and would have to dig its way out. A similar process occurs at the center of the sun but in reverse. The protons in the sun fuse together and then release the energy that causes the sun to shine. Even though the core temperature of the sun isn't high enough for it, the protons manage to come out through the energy barrier.

In technical terms, quantum theory can be defined as the theory of objects that are isolated from their surroundings. It's not very easy to isolate these large objects from their surroundings, so it turns into a theory of the microscopic world of atoms as well as the subatomic particles. This fact is especially true for certain parts of the theory which depend upon the exact indistinguishability of fundamental particles, something that is impossible to come across in the world of large-scale, composite objects.

One of the greatest physicists of all times, Richard Feynman, who has won the Nobel Prize for Physics had stated, *"I think I can safely say that nobody understands quantum mechanics"*. On the other hand, one of the pioneers of the quantum theory, Neils Bohr claimed that: *"Anyone who is not shocked by <u>quantum theory</u> has not understood it"*.

Contrary to popular belief, a lot of research has been done with regards to classical physics. In fact, studies have been conducted since the early 1 9 0 0 ' s . B a c k   i n   t h e   1 9 2 0 s   a n d 1930s, Schrödinger, Bohr, Heisenberg, and many others had discovered the fact that the atomic world is filled with chaos and murkiness, and it's not exactly a precision clockwork as depicted by the classical theory. Classical physics can in fact be considered one of the closest proximations to quantum physics especially under conditions where a large number of particles are involved. Of course, the classical physics had conducted some in-depth research until the 20th Century and the study continues to serve us for the majority of our everyday purposes.

Modern physics, which includes general relativity as well as quantum physics, is more fundamental, all-encompassing, and overall a more accurate physics which has been taken to a different level. For instance, position and momentum are actually the approximations of the world containing larger-sized things and is often referred to as the classical world. But in reality, the hidden quantum world is very different. For people who don't understand it fully, the quantum world can remain a perpetual enigma.

As we have seen above, even the most basic understanding of the quantum theory necds a little background explanation of atomic theory while the certain parts of the uncertainty principle require an in-depth discussion before you are in a position to make any sense of it all. It is only then that we can understand the most bizarre and obscure aspects of the quantum theory. This includes superposition, wave-particle, decoherence, non-locality etc.

Chapter Eight: Quantum Entanglement

One of the central principles of quantum physics is Quantum Entanglement. It is a physical phenomenon that occurs when pairs or else groups of particles are generated, they interact or share spatial nearness in a way that we would not be able to describe the quantum state of each particle independently of the others. This is even when these particles are separated from each other by a large distance. Here, the quantum state has to be described for the whole system. This means that the measurement of the quantum state of one particle would determine the possible quantum state of the other particles as well. The connection between them is not dependent on their location in space. Quantum entanglement may seem to instantly transmit information; there is no movement through space, so it does not violate the speed of light.

An entangled system is one whose quantum state is not factored as the product of the state of its constituents. In an entangled system, the particles don't exist individually but are inseparable. Either of the constituents of the system cannot be completely comprehended without consideration of the other constituent.

Various types of interactions can cause quantum systems to become entangled. This entanglement is broken when there is decoherence due to interaction with the surrounding environment. The paradox of measuring either of the particles is that it will cause the collapse of the entire entangled system instantly. So, it would be impossible for any information to be communicated to the other particle from the other and thus assure proper outcome when one is measured.

Usually, entanglement is created when there are direct interactions between any subatomic particles. Such interaction may take various forms. Spontaneous parametric down conversion is one of the commonly used methods to generate photon pairs that are entangled in polarization. Another method might use a fiber coupler to confine the photons and mix them. When the early tests for Bells theorem were carried out, atomic cascades were used to generate entangled particles. Even without direct interaction, entanglement between quantum systems can occur. One of these methods would be through entanglement swapping. Sometimes systems are naturally entangled; multi-electron atoms always have entangled electrons in their electron shells. Electron entanglement needs to be considered in order to correctly calculate the ionization energy for such atoms.

Measurement of any physical property like momentum, position, spin, etc. of entangled properties is found to be correlated. When such measurements are attempted, it appears as if there is a paradox i.e., the measurement of any property of a particle will cause an irreversible collapse on the particle and change its original quantum state. In entangled particles, such measurements will be of the whole system. In entangled pairs, it appears as if one particle "knows" what measurement was performed on the other and what the outcome was. This happens even though there is no possible way for this kind of information to be exchanged amongst the particles in the pair, especially when they are separated from each other by huge distances.

This kind of phenomenon was the subject of Albert Einstein, Boris Podolsky, and Nathan Rose's 1935 paper. Later, Erwin Schrodinger also wrote several papers on this and it later came to be known as the EPR paradox. The EPR paradox is actually a classic example of quantum entanglement and we will explain it further.

Bohr and Einstein had a debate about the

fundamental nature of reality. The core of this debate was the question about if there was one objective physical reality that every observer saw from their own viewpoint, or if the observer creates physical reality with the questions he posed with experiments. The former was Einstein's view and the latter were Bohr's. Until his death, Einstein struggled with the view that no objective physical reality other than the one that is revealed through measurement from quantum mechanical formulation existed. Over the years, more studies have been carried out and most of these agree with Bohr's view rather than Einstein's.

Amongst the modern resolutions, one is as follows; when two entangled particles are created at the same time, their measurable properties have proper defined meanings just for that ensemble system. The properties of individual subsystems would remain undefined.

If corresponding measurements are carried out on the entangled subsystems, a correlation between the outcomes will always be there and there will exist a properly defined outcome for the ensemble. If you consider the outcomes for each subsystem separately at every repetition, the measurements will not be well defined. It will also be hard to predict these outcomes. This resolution removes any need for hidden variables or other schemes that were introduced over the years in order to explain quantum entanglement.

In quantum physics, a pair of quantum systems might be described as a single wave function under some conditions. This wave function would encode the probabilities of the various outcomes that would result from any experiments performed on the two quantum systems, either individually or even jointly. At the same time, it was also written that sometimes the outcome of an experiment couldn't be predicted. Such indeterminacy is visible when a beam of light falls on a mirror that is half-silvered. One half of the beam of light will be reflected while the other half will pass. Quantum mechanics cannot be used to predict if the light would reflect or pass, if that beam was reduced to a single photon that falls on the mirror.

The Einstein-Podolsky-Rosen paradox

Its namesakes Albert Einstein, Boris Podolsky,

and Nathan Rosen proposed the EPR paradox or Einstein-Podolsky-Rosen paradox. Like the Schrodinger's cat experiment, this was also a thought experiment. Scientists like Einstein considered the behavior of quantum entanglement impossible. This is because it violated the realist views and so they argued that the present quantum mechanics formulation was probably incomplete. They attempted to show how wave function did not provide complete realistic information and that the Copenhagen interpretation was unsatisfactory. This work was carried out after Einstein joined Princeton University, at the Institute of Advanced Study. This was after he had escaped Nazi Germany, in 1934.

For a simple version of EPR Paradox, consider the following case. An original particle exists with quantum spin 0 and it decays into two new particles. There is a Particle X and Particle Y which head off in opposite directions from each other. The original particle has a quantum speed of 0. Both the particles X and Y have quantum spin ½. They have to add to 0 so they are +1/2 and -1/2 each. From this relationship we see that these particles are entangled with each other. Measuring Particle X's spin will have an impact on the possible result of measuring Particle Y's spin. This is not just a theoretical prediction but a result that has been verified through experiments of Bell's Theorem.

Heisenberg's uncertainty principle provided a routine explanation for this at that time. Conjugate pairs exist such as position and momentum, time and energy, etc. Physical quantities exist in pairs as conjugate quantities. If one is measured, the other conjugate quantity becomes indeterminate. According to Heisenberg, this uncertainty is due to quantization of disturbance from measurement.

In 1935, when the EPR paper was written, they intended to show that this explanation was not sufficient. They considered two particles A and B that were entangled and pointed out that the measurement of Particle A would cause a quantity of Conjugate Particle B to become indeterminate even without any contact or classical disturbance. The point was that quantum states of the two particles of a system could not always be decomposed from their joint state. This was the case in Bell's state as well. Heisenberg had attempted to provide an explanation for non-locality, which was a quantum effect. EPR suggests two possible clarifications, either there was an interaction despite being separated or else information about a possible outcome was already present in each of the particles.

The second explanation suggests that information about the outcome of any possible measurements were already embedded in the two entangled particles. This was the preferred explanation of EPR, since its authors believed in the hidden parameters that encoded this information. The theory of relativity causes the first suggestion to be conflicting. The conclusion was that quantum mechanics was still be incomplete since hidden parameters were not considered in its formulations.

If we talk about the Copenhagen Interpretation that is explained elsewhere in this text, Albert Einstein was one of the most prominent opposers of it. Quantum mechanics in his view was incomplete. Other writers offered the suggestion that there had to be hidden variables that were responsible for random measurement results. The EPR paper in 1935 made it a physical argument. The authors of this paper claim that for a specific experiment whose outcome is known before the measurement takes place, there has to be an element of reality or something that exists to determine the measurement outcome. According to them, such elements or reality are local, they each belong to a specific point in space-time. Each element could only be influenced by any event that is located in the past. These claims were founded on local realism.

The EPR paper

The original EPR paper was actually primarily authored by Podolsky based on discussions with Einstein and Rosen. The paper challenged the prediction of quantum physics that suggested that it was impossible to know the position and momentum of a quantum particle at the same time. The paper tried to show what would happen to two systems that were allowed to interact and after a while it is assumed there is no interaction any longer. It was explained that two particles were allowed to interact briefly and then they moved off in the opposite direction to each other. According to Heisenberg's uncertainty principle, measuring the momentum and position of the second particle would be impossible, but it is possible to find the exact position for the first particle. You can deduce the measurements of the position of the second particle through some calculations. Similarly, if the momentum for the first article can be measured, it can be calculated for the second particle as well. The EPR paper argued that they could thus find the exact values or position and momentum of the second particle. They deduced that the second particle thus has a real position and a real momentum. EPR thought that the experiment tried to show that although quantum theory says that both values of a particle cannot be known, the former says that determinate values for each must exist. They concluded that the quantum mechanical description of physical reality that wave functions give is incomplete.

The actual EPR paper was concluded with the following words:

"While we have thus shown that the wave function does not provide a complete description of the physical reality, we left open the question of whether or not such a description exists. We believe, however, that such a theory is possible."

Bell's Inequality

Bell's theorem was named after John Stewart Bell and draws a very important distinction between quantum mechanics and classical mechanics. It particularly pays attention to quantum entanglement where particles in a quantum state stay mutually dependent even after they are physically separated. Many experiments were carried out to verify Bell's theorem and it proved that quantum entanglement exists even across large distances. For applications of quantum physics like quantum computing, this quantum entanglement plays a major role.

Bell's theorem stated that no physical theory of local hidden variables could ever reproduce all the predictions of quantum mechanics. This theorem has played an important role in Physics. While some appraised it, others utterly ridiculed the theorem. It rules out any hidden variables as a feasible explanation of quantum mechanics. He also used superdeterminism to address this theorem in 1985 and it was very unpopular at the time. One of the important parts of Bell's work was that he made a deliberate effort to encourage work on any completeness issues that had fallen into disrepute. He actually derived the theorem in his 1964 paper, which referred to the 1935 paper of Einstein, Podolsky, and Rosen. He wrote of the same two paradoxes that were referred to in that paper and from those he derived Bell's inequality. This inequality's violation is that one or the other of the above-mentioned assumptions has to be untrue. Unlike the EPR paper Bell's paper took hidden variables into consideration and not just reality, and another advantage was that Bell's inequality was a testable experiment. Later it was generalized that his theorem was not just about hidden theorems but more about any outcomes that might have taken place other than the one that actually did. The existence of these is known as assumption of counterfactual definiteness or assumption of realism.

The inequality derived by Bell is written as follows:

Here,

P, b and c represent three arbitrary setting of the two analyzers.

The theory of hidden variables fails when we consider measurements of the spin particles along different axes. If many such pairs are made for measurements and if the hidden variable's view is correct, then the results should always be able to satisfy the theory of Bell's inequality.

Many experiments have shown that Bell's inequality is actually not satisfied. The main issue about measuring along different axes is that the measurements could not have exact values at the same time. These are incompatible because the maximum simultaneous precision of such measurements is restrained by Heisenberg's uncertainty principle. It is in contrast to classical physics.

Entanglement has been shown to be a fundamentally non-classical phenomenon and mathematics has proved that compatible measurements would not result in Bell's inequality-violating correlations. There are also two classes of Bell's inequalities to consider. Clauser and Horne distinguished Bell inequalities as homogeneous or HBI and Inhomogeneous or IBI.

Classification of Entanglement

All quantum states are not equally valuable as resources. Different entanglement measures are used to quantify this value, and these assign each quantum state with a numerical value. It is usually more interesting to settle for a rougher way when comparing quantum states. These give rise to various classification schemes. The fewer the allowed set of classifications, the finer the classification will be. One example is – when two states can be altered into each other using a local unitary operation, they belong in the same LU class. This class is considered the finest of the classes generally used. If two states are together in the same LU class, then they have the same values for entanglement measurement and also the same value in distant lab setting as a resource. An infinite number of LU classes exist.

Applications of quantum entanglement

Quantum entanglement is used for many things in quantum information theory. Some tasks that would otherwise be impossible are made possible with entanglement. Some of the best-known entanglement applications include quantum teleportation and super dense coding. It is also used in the protocols sometimes for quantum cryptography since the shared noise is useful as a one-time pad. Quantum cryptography that uses entanglement allows the sender and receiver to easily detect an interceptor during their communication. Entanglement is also used in interferometry.

Chapter Nine: Schrodinger's Cat & Schrodinger Equation.

Erwin Schrodinger was an Austrian Physicist who postulated this thought experiment in 1935. Schrodinger's cat is sometimes called a paradox but is very famous to this day. In this experiment, the physicist describes a scenario where a cat may be both dead or alive. To explain further, let us look at the experiment. In this, a cat is kept in a box with a flask of poison and a radioactive source. If a single atom decayed, the flask is broken, the poison will be released, and the cat will die.

The Copenhagen interpretation of quantum physics suggests that the cat can be both dead and alive and you won't know until the box is open. They call this a state of quantum superposition where the object can exist in multiple states, which could lead to different outcomes. Schrodinger wanted to show how the scenario that the cat could be both dead and alive at the same time, is absurd. According to him, the cat either had to be dead or be alive. That was his view, a lot of scientists to date agree with the Copenhagen interpretation.

The Schrodinger's cat experiment is still used to compare the pros and cons of any interpretation in quantum physics. Schrodinger's experiment posed the question of when a quantum system stopped existing in the state of superposition and starts existing as one of them. The role of the observer is still a very important question in quantum physics due to this experiment. There is a lot of speculation and conjecture about this in pop culture as well as quantum computing. The experiment is actually a purely theoretical one that Schrodinger proposed and not one he physically carried out. The apparatus is not known to be constructed. Similar experiments have been performed over the years.

Out of these, a prominent one was "Winger's friend", which was a variant of the Schrodinger's cat experiment but having two observers. In this, it is questioned if the wave function collapses when any one observer looks into the box, or if it collapses only when he tells the other person what he observes. Many successful experiments that involved similar principles were conducted.

Let us consider some other interpretations of quantum physics, related to Schrodinger's cat, that have been proposed.

Copenhagen Interpretation:

The Copenhagen interpretation is one of the more commonly considered ones. In this, they say that the state of superposition stops existing and an observation is made. Then there is only one or either of the states that were earlier in superposition. According to the Copenhagen interpretation, the cat experiment can be made sense of by saying that the cat is alive with the nucleus un-decayed and the cat is dead with the decayed nucleus. This state of superposition stops existing only when you open the box to check.

Many Worlds Interpretation:

Another interpretation is the many worlds interpretation suggested by Hugh Everett in 1957. He does not place importance on observation as a crucial part. According to this interpretation, even after the box is opened, the cat can be in the superposition states of its existence, but these are decoherent from each other. It means that if the box is opened, there are two scenarios. One is where the observer sees the dead cat and one where the cat is still alive. These states are decoherent so there is no interaction between them. Two different observer states are formed which are linked to the cat, so the observation and the actual state of the cat correspond with each other. Due to quantum decoherence, different outcomes will not have any effective communication amongst each other. Quantum decoherence is also important with regards to consistent history. The dead cat or the alive cat is the only part of this interpretation that can be a part of consistent history.

Ensemble Interpretation:

The ensemble interpretation actually just considers Schrodinger's cat a paradox and nothing of importance. According to the interpretation, superpositions are just sub ensembles of a large statistical ensemble. It states that the state vector would apply to many cat experiments that are similar but not to any individual cat experiment.

Relational Interpretation:

The relational interpretation states that there is no fundamental difference between the cat, the observer, or the apparatus or any animate or inanimate system. All of these are quantum systems where the exact same rules of wavelength evolution will apply, and they are all observers. This interpretation also shows that different observers will have different views about the system and it will depend on all that they know about it. This interpretation states how the cat is an observer of the apparatus; the human observes the experiment as a whole. Before the human opens the box, he doesn't know of the state inside, but the cat does. So automatically, at the same moment, both observers have different amounts of information and thus observation. The state of the system collapses into one only when the box will be opened.

Transactional Interpretation:

Transactional interpretation has apparatus that

emits advanced waves backward in time and this combines with the wave, which is emitted forward in time by the source. Together, the backward and forward waveform a standing wave. These waves are considered real physically, and the apparatus is the observer here.

In this interpretation, the wave function collapses and this is atemporal and occurs during the whole transaction amongst the apparatus and source. Here the cat is never in a state of superposition. It is always only in one state at a particular moment in time no matter when the observer looks inside the box. The quantum paradox is resolved by the transactional interpretation.

Zeno Effect

The Zeno effect causes delays in any changes from the initial state. On the other hand, the anti-Zeno effect increases the rate at which these changes take place. If you consider this for the cat experiment, there would be two scenarios. The observer could keep looking into the box and this would either cause the final state to happen faster or cause a delay in it. The Zeno and Anti-Zeno effect are considered very real and have been known to occur to actual atoms. The quantum system and the surroundings must be closely coupled together to get accurate information. When there is no information available to the observer about what is inside the box, it would be a quasi-measurement. When the observer looks inside the box, only then can an actual measurement be taken. Quasi-measurements are a cause of the Zeno effect. This teaches us that the death of the cat would be delayed or accelerated in accordance to its environment and not by the observer looking into it.

Objective Collapse Theories:

The objective collapse theories state that superpositions will be spontaneously destroyed when an objective physical threshold will be reached. This would happen irrespective of any external observation. The physical thresholds could be time, temperature, mass, etc. The cat is therefore expected to already be settled into its final state before the observer opens the box. These theories need a modification in standard quantum mechanical theories so that superpositions can be destroyed by the process of evolution of time.

Schrodinger was actually a major contributor to the field of quantum physics. He developed many fundamental results of quantum theory, which became a base for wave mechanics. Erwin Schrodinger also shared a Nobel Prize for his work in 1933 with P.A.M. Dirac. He took the Bohr atom model a step further by using mathematical equations to describe how likely it was to find an electron in a particular position. This is known as the quantum mechanical model of an atom. This model predicts the probability of the location of electrons and does not define its actual path, unlike the Bohr model. He also formulated an equation known as the Schrodinger equation, which is the fundamental equation in the field of quantum mechanics. It is as important in this field as Newton's laws of motion are in classical physics.

Schrodinger Equation

The Schrodinger equation is essentially a wave equation that is used to define the changes of a physical scheme over time, where quantum consequences like wave-particle duality are significant. Such systems are called quantum systems. The derivation of the Schrodinger equation was a landmark in developing quantum theory. Schrodinger applied the equation to the Hydrogen atom and accurately predicted many properties. This established the precision of this equation. The equation is widely used in physics related to atoms, solid states, and nuclear field.

Like I said, the form of the equation will depend on the physical situations it is under. The equation is in terms of the wave function of the particle. Once you solve it, you will know what the wave function will look like for the particle at some point in the future and so we can determine distribution of the position, momentum, etc. of the particle in future. Position and momentum in classical mechanics is solved as functions of time. In quantum mechanics, you need to solve for a wave function. This wave function is already a function of 3 coordinates as a function of time. The equation is a partial differential equation. Newton's second law isn't enough to find out the future motion of the particle. The formula for F force is required.

This is because for one, the particle's motion will be different in different fields depending on the different forces exerted on it. Schrodinger's equation itself is also not enough for this. It is also important to know the nature of the system.

For a time-dependent equation, Schrodinger equation looks as below:

Here,

I= imaginary unit

ħ = reduced Planck constant

D/dt= derivative with respect to time

Psi= state vector of quantum system

$\hat{H}$ = Hamiltonian operator

For a time independent equation, Schrodinger equation looks as below:

$$\hat{H}|\Psi> = E|\Psi>$$

Here,

E= Constant equal to total energy of the system

This form is used only when the Hamiltonian operator is not dependent on time.

When you study these equations in further detail, you might think that the equation is like a tautology. The equation itself does not have any predictive power unless you already know what H is or have an idea about the form it could take. The equation might even just be a type of definition for the Hamiltonian operator. In general, the Schrodinger equation will state that when the kinetic momentum of the particle is squared and divided by twice the mass of that particle, it equals the kinetic energy of that particle.

Limitations of Schrodinger equation:

- It cannot describe the systems in which the number of particles will vary.

- It cannot describe the systems in which the particles are relativistic.

- It cannot describe particles with spin.

Chapter Ten: Applications of Quantum Physics:

Quantum physics has had many valid applications in real life. We will explain some of them to you here:

In Computers and Smartphones and Other Electronics

The Principles of quantum physics is what the entire computer industry stands on. It is because of quantum physics that we know that the electrical properties of silicon can be manipulated due to the wave nature of electrons. The study of semiconductors was also crucial for this. It led to the invention of the transistor and diode, which are essential electronic parts. Computer chips that are used in our daily gadgets were made by using the principles of quantum physics as well. So basically, any laptops, desktop, phone, tablet, etc. that you use are powered by computer chips and these would not have existed without quantum physics. Even your light switch works with the effect of quantum tunneling. Quantum tunneling is also used in USB drives when they erase their memory.

Cryptography

Cryptography is the art of encrypting and sending a message and decrypting it for the receiver. It requires an algorithm that can be used to mix the message while sending it and for a key to be able to simplify it again. A secure cryptosystem should be impossible to crack by anyone other than the intended recipient or authorized party. One of the main issues in cryptography is key distribution. Classical cryptography uses private key description of RSA type algorithms. Modern cryptography is much safer than classical cryptography as it allows guaranteed security in transmission of information. It uses trapdoor algorithm for the public key. The complexity of the computations should make it impossible to decipher the message with most computers; however, if there were major breakthroughs that make advanced computers available to the public, it would be easy to crack the code. Quantum physics does not use any specific mathematical algorithm that can be easily cracked. This is why a quantum cryptosystem is being developed to make the process completely secure. There is still a lot of work being done to fully develop quantum cryptography. Quantum physics has a positive advantage in this field with the help of the EPR protocol and Bell measurements. Another advantage of it is that it allows the detection of eavesdropping easily.

Quantum bits have a natural behavior that is explained by the observer effect. Due to this, if any bit were in a superposition state, it would collapse into eigenstate. The intended recipient would know when a bit was no longer secure. Although confidentiality was the main objective for cryptography, it has different applications as well such as authentication and for digital signatures. The concept of quantum cryptography has the chance of being the ideal system of secure encryption of data transfer.

Lasers and Telecommunication

The principles on which fiber optic telecommunications are based all rely on quantum physics. Lasers are used to send messages down fiber optic cables, regardless of the fibers. Here the light sources are lasers that are quantum devices. Practical use of quantum mechanics is seen in the simplest of functions such as when the UPC labels of groceries are scanned using a laser. Even when you make a call you are using quantum physics.

Quantum Voltage Standard

Quantum theory has also been used to establish a voltage standard that has proven to be extremely accurate. When two layers of superconducting materials are separated from each other with a thin insulating barrier, a supercurrent may pass from one to the other. The supercurrent is just a current of paired electrons that pass from one superconductor to the other superconductor. It is an example of the tunneling process as we have mentioned before. Brian D. Josephson was a British physicist who described most of the effects in 1962. These are now referred to as Josephson effects and have been demonstrated experimentally as well.

We get the formula 2eV=hv from Planck's relationship here because the energy of the electron pair becomes 2eV when it crosses the junction. This is if a direct current voltage (V) is applied across the superconductors.

The oscillatory behavior of such a supercurrent is called the alternating current Josephson effect. Even though this has been directly detected, it is very weak. Studying effects that result from the supercurrent interacting with microwave radiation will be a more sensitive investigative method.

MRI or Magnetic Resonance Imaging

An MRI test sends images of internal organs and structures to a screen using a magnetic field and radio wave energy. We all know what a huge role MRI's have played in the medical field. It is a major advantage in being able to detect serious medical conditions. An electron's spin is relatively oriented to the nucleus's spin and this causes a shift in energy like in atomic clocks.

This causes the electrons, protons, and neutrons to behave like tiny magnets and is responsible for the application of quantum physics in Magnetic Resonance Imaging. The central process in this is called Nuclear Magnetic Resonance.

Advanced Microscopes

A research team at Hokkaido University in Japan developed the world's first entanglement enhanced microscope. This microscope blazes 2 beams of photons towards the substance and processes the pattern that the reflected beams generate. Depending on if the surface is flat or uneven, the pattern might differ. The same principle is used to improve resolution in tools of astronomy like interferometers. The different waves of light will superimpose and give us better results for analysis.

GPS

These days nearly all of us use a GPS server on our phones or cars to find locations. This Global Positioning System is actually dependent on the application of quantum physics. It is a satellite network that broadcasts time and enables navigation. Each satellite has an ensemble of the atomic clocks we mentioned earlier that depend on quantum mechanics. So, every time you use your GPS server to navigate directions, you are using quantum physics again.

X-rays

X-rays are a form of electromagnetic radiation that has wavelengths between 0.01nm and 10nm. A vacuum tube, particle accelerator, or an x-ray tube can produce these. These would not be possible without quantum mechanics.

Cesium Clock

The most accurate type of clock invented to date is the Cesium clock. Transitions between the spin states of Cesium nucleus are used to produce a regular frequency that has been utilized for establishing the time standard of the world. Atomic nuclei have spin similar to electrons. When these nuclei spin, a set of small effects are produced in the spectra. These are known as hyperfine structure. These effects in the spectra are small due to the fact that the magnetic moment is relatively small even though the spinning nucleus has the same magnitude as an electron. The magnetic moment is what governs the atomic level energies. The spin quantum number of the Cesium atom is 7/2. The spin angular momentum of the single valence electron is combined with that of the nucleus to obtain the total angular momentum of its lowest energy states. The angular momentum of all the other electrons adds up to zero, other than the valence electron. Only the spin angular momenta are considered because all the ground states have zero orbital momenta. If the nuclear spin is considered, then the total angular momentum will usually be characterized by a quantum number that is denoted by "F" for Cesium, the F value is 3 or 4.

Elitzur-Vaidman Bomb Tester

This bomb-testing problem was proposed in 1993 by Avshalom Elitzur and Lev Vaidman. It was then successfully constructed and tested in 1994 by Anton Zeilinger by using the Mach Zehnder interferometer. It is called interaction-free measurement. There is a simple way to explain this bomb-testing problem. Think of a collection of bombs in which a few are duds that won't work. Single photons can be used to detonate these bombs. A good bomb will absorb the photon and explode. A dud bomb will not absorb the photon and hence cannot explode. The counterfactual phenomenon of quantum physics can be used to separate dud bombs from usable ones. Testing by detonating bombs will just lead to the destruction of the useful ones. The plunger can activate the detonator of a bomb when it is attached to an extremely sensitive mirror. A photon impinging on it would push the plunger and cause detonation. In duds, these plungers are stuck and so they cannot get pushed or activated. So, we see that dud bombs reflect the photons. If the photon doesn't hit the mirror of the bomb, it tells us that it went through the null measurement path. A Mach Zehnder interferometer was used in the experiment where a low intensity light source was used so that only one photon was emitted at a time. This photon has equal chances of getting reflected or passing through the beam splitter BSI. If the bomb is good, then the photon is absorbed and if it is a dud bomb the photon passes through unaffected. The test will identify a third of the good bombs without detonating, but the remaining two-thirds of the good ones are detonated. In 1996 Kwiat et al invented a technique that uses polarizing devices to yield an arbitrarily close rate to one. This would determine the answers without setting off the bombs.

Fluorescence and Phosphorescence

Fluorescence is a type of photoluminescence that is emitted by a substance that has absorbed light or some other form of electromagnetic radiation. This light that is emitted will have a longer wavelength and lower energy than the radiation that was absorbed by the substance. For shorter wavelength, two photons need to be absorbed by one electron from the absorbed radiation. Glow in the dark paint, toys, etc. are examples of fluorescence or phosphorescence. For instance, some clock dials also glow at night when they have absorbed enough light charge during the day.

Quantum teleportation

One of the modern applications of quantum physics is the possibility of exploiting quantum entanglement for quantum teleportation. The proof of teleportation in Zeilinger experiment shows that the nature of quantum entanglement is crucial. To understand this, we can consider two people, Susan and John, who share a pair of photons. These photons were prepared in an entangled Bell state of polarization. We see that each of them possesses an entangled photon. Susan also possesses a photon that is in an unknown polarization state. Now Susan shall take measurements of the photons that are in her possession. This will have the possibility of four outcomes. John's photon could transform into any of these four states. It would depend on the outcomes of Susan's measurements. Susan's operation causes entanglement of the two photons in her possession and John's photon gets disentangled. Susan now informs John of the outcome of her measurement and John sees that the disentangled state of his photon is equal to the unknown polarization state of Susan's second photon, or it is equal to it by a local operation. We can consider that Susan's photon was teleported to John since the quantum state of the photon was transferred from her to him. Non-local quantum correlation is clearly demonstrated by this phenomenon.

As you can see, quantum physics is quite essential to our daily lives in the modern age. It is used from the simplest thing like determine the time using Cesium clocks to more important matters like diagnosing diseases using MRI's. All the applications we mentioned above and many more are all dependent on the principles of quantum physics. All these just highlight the importance of studying this field even further and in depth in the future.

Chapter Eleven: Superconductivity

Superconductivity is a quantum mechanical phenomenon. It is the complete desertion of electrical resistance magnetic flux fields in certain materials when they are cooled below a characteristic temperature. Such materials are called superconductors.

The phenomenon of superconductivity was discovered by Heike Kamerlingh Onnes in 1911. He found that at a temperature of around 4.2 °K, the resistance of mercury suddenly drops to zero. This discovery allowed him to win the Nobel Prize in 1913. This was followed by numerous studies over years to find other materials which are superconducting at absolute zero temperature. In 1933, R.S Ochsenfeld and W. Meissner found that if a metal is cooled into its superconducting state in a magnetic field, it expels the field from the interior. In 1986 there was another major breakthrough when Karl Alexander Muller and J. George Bednorz discovered a class of copper oxide materials which become superconductors at temperatures above 70°K. This earned them the Nobel Prize in 1987 for physics. A Japanese team with Hideo Hosono discovered an iron arsenic high-temperature superconductor in 2008. Since the discovery of these, many such superconductors have been discovered. Metals like zinc, magnesium, aluminum, mercury, and tin etc. display characteristics of classical superconductivity. Metals like molybdenum display superconductivity only after high purification. Many such elements have shown to be superconducting at near absolute zero temperature. In certain cases, like Europium, extreme pressure also needs to be added for this to act as a superconductor.

The critical temperature is the temperature at which the electrical resistance of the material drops to zero. This sudden transition appears as though a new state of matter is created. Usually most materials exhibit this superconducting state at low temperatures. Until 1986, the highest known critical temperature was 23k. Then later, some high-temperature superconducting metals were discovered. The ones that have a critical temperature above 120K usually get a lot of attention, as it is possible to maintain this state with liquid nitrogen at 77K.

The BCS theory of classical superconductivity was proposed by Leon Cooper, John Schrieffer, and John Bardeen, for which they won the Nobel

Prize in the year 1972. Their theory proposes that electrons in an electric current move in pairs at very low temperatures. These pairs are called Cooper Pairs. This allows them to move across a crystal lattice without collisions disturbing their motion. It also says that the energy of the interaction is quite weak between the electrons of the pairs are easily broken by thermal energy. This is the reason superconductivity occurs at such low temperatures. The theory does not account for the existence of superconductors of high-temperatures around 80k. For these, other mechanisms are invoked for an explanation. The BCS theory is still considered standard for explaining superconductivity. The Meissner effect, as we now know it, was named after W. Meissner who was the first one to observe it in 1933. Below a critical temperature, materials tend to undergo a transition into the state of superconducting. It is characterized by two properties at this point. One is that they don't offer any resistance to the passage of electrical current. As the resistance falls to zero, inside the material, a current can circulate without dissipation of any energy. The second is that if they are sufficiently weak, no external magnetic field could penetrate the superconductor and will just remain at the surface. This phenomenon of field expulsion is the Meissner effect.

The state of superconducting may be destroyed with a rise in temperature or the magnetic field which would then penetrate inside and suppress the Meissner effect. This perspective allows there to be a distinction between two superconductor types. Type I superconductors remain in the superconducting state only when the applied relative magnetic field is weak. The BCS theory explains Type I conductors very well. Type II material remain as superconductors even when there is local penetration of a magnetic field and this allows them to preserve their properties of superconducting. They are usually alloys of different metals. This type of behavior is explained by the existence of a state where superconductive and non-superconductive areas coexist in the material. This mixed state is sometimes called the vortex state. This is why we use Type II superconductors in high magnetic fields for things like particle accelerator magnets. On the other hand, Type I superconductors have not been of much practical use. This is because of the small critical magnetic fields and that there is sudden disappearance of the superconducting state at that temperature. Alloys like lead-bismuth are Type II superconductors. Metals like Zinc and Mercury are Type I superconductors. Referring to their properties, Type I are referred to as soft superconductors and Type II are called hard superconductors.

Superconducting can be utilized for various purposes like energy storage systems, motors, generators, medical imaging devices, computer parts, etc. The advantage of any devices made with superconducting materials is that they have low power dissipation, high sensitivity, and high-speed operation. Superconductive electronic devices can be formed from superconductors and films of normal metals. These can replace transistors in certain applications. It has also been observed that the current flow continues in a superconducting circuit even after the source has been cut off. Due to this property, it has been possible to create powerful electromagnets that retain magnetism indefinitely once they are energized.

Such powerful superconducting electromagnets are very useful in magnetic resonance imaging. Ultrasensitive superconducting circuits have been made to study the heart and brain in humans, and also for many other scientific experiments.

Chapter Twelve: Quantum Computing

Quantum computing uses quantum mechanics to develop the computing field. It uses phenomena like superposition and entanglement for this field. A device that uses quantum computing is known as a quantum computer.

The development of quantum computers is a major goal of the industry. It is expected to perform certain tasks exponentially faster than standard computers. Quantum computers are much more complex than the standard computers we use in regular life. Standard computers will encode information in a string of binary digits or as bits 1 or 0. Quantum computers use quantum bits also known as qubits. These qubits can be both 1 and 0 at the same time, as they exist in a superposition of states. The quantum programmer can manipulate the superposition of quantum bits to solve problems that the standard computer effectively cannot. Quantum computing in the future might help in a massive advancement in the fields of logistics, finance, medicine, artificial intelligence, etc.

A quantum Turing machine, also known as the universal quantum computer, is a theoretical model of this type of computer. Paul Benioff and Yuri Manin were the first ones to initiate the field of quantum computers in 1980 and then in 1982 it was Richard Feynman and David Deutsch in the year 1985. A lot of work has been done while researching this field, it is still in the infant stage. Quantum computers would be able to deal with problems on a much larger scale than standard computers. For instance, instead of the algorithms used by regular computers now, quantum devices would use advanced ones like Simon's quantum algorithm. This algorithm runs faster than any known algorithm right now. Practical as well as theoretical research is ongoing as a lot of funding is provided to develop this field. Quantum computers will help in many fields like national security, business, trade, etc. The IBM Quantum Experience has a 20-qubit computer available for use. D-Wave Systems are developing a new version as well.

Richard Feynman was a major pioneer in the field of quantum computing. In 1959 he stated the possibilities of using quantum physics in computing in his lecture. Then in 1981 he placed great emphasis on promoting the development of quantum computers at an MIT and IBM conference. In 1993, a group of international physicists showed the possibility of quantum teleportation. Peter Shor at Bell Labs discovered an important quantum algorithm in 1994.

Compared to standard computers, this allowed quantum computers to work on large integers exponentially faster. This was one of the major breakthroughs in the field. David P. DiVincenzo's paper on "The Physical Implementation of Quantum Computation" was another important milestone. In 2005 a semiconductor chip ion trap was built by researchers at the University of Michigan. In 2009 the first solid-state quantum processor was built at Yale.

In 2010 there were further developments on chips based on quantum optics and digital combination circuits were designed using Symmetric Functions from quantum gates. Another breakthrough was when a team from Japan and Australia transferred a set of complex quantum data with full transmission energy and managed not to affect the qubits superpositions. The D-Wave One was announced in 2011. The same year, it was also proved that Von Neumann architecture could be used to make quantum computers. In 2012 a university in Hefei, China, reported the first successful quantum teleportation from one to another macroscopic object. A new technique called boson sampling was reported in 2013. The 2013 launch of the Quantum Artificial Intelligence Lab was announced by Google. The U.S. National

Security Agency also runs a multi-million-dollar program to develop a quantum device to break encryption.

A quantum logic gate in silicon was built for the first time in 2015 at the University of New South Wales. NASA unveiled the first fully functional quantum computer at the end of 2015. It was made by D-Wave, a Canadian company. So far only two of these devices have been made. The IBM Quantum Experience allows quantum computing to be available for the public via the cloud. Their processor is made of 5 superconducting qubits and housed at IBM T.J. Watson Research Center in New York. In 2016, scientists at the University of Maryland built the first reprogrammable quantum device. In 2017 IBM announced that they would make quantum computers available commercially and these would be known as IBM Q. Then in May 2017, a group of researchers in the US announced that they built a 51-qubit-quantum simulator. Mikhail Lukin announced this at a conference in Moscow and this simulator was designed to solve a single equation. A new system would be required for a different equation. In 2018 scientists announced that they discovered a new form of light, which involved polaritons that would be useful in quantum computing. The first multi-qubit demonstration was performed on a trapped ion system in a quantum chemistry calculation by a team led by University of Sydney in July 2018.

As you can see, over decades a lot of milestones have been reached in quantum computing since it was first introduced. Let's get a better understanding of Quantum Computing:

Basically, a classical computer's memory is made of bits that are each represented by one or zero. On the other hand, a quantum computer has a sequence of qubits which could represent one, zero or any quantum superposition of them. That means that a quantum computer with qubits "n" could be in arbitrary superposition of $2n$ different states simultaneously. The device uses quantum gates and measurement for operations. Algorithms have a fixed quantum logic gate sequence. The problems are encoded similar to standard computers but by using qubits. When the calculation is ending it will usually have a $2n$ eigenstate where the system of qubits collapse. Quantum algorithms will usually only provide a solution with a known probability. They are probabilistic but not non-deterministic.

If a quantum computer and standard computer have the same number of qubits and bits respectively, they are still fundamentally different from each other. If you want to represent an n qubit system in a standard computer, it needs to store 2n complex coefficients. If it is an n bit system, then having the numerical values of the n bits is enough. You need to remember that although qubit computers may seem to hold exponentially more information, the qubits are still just in a probabilistic superposition state. So, when the final state will be measured, only one possible configuration will be found. The fact that qubits are in a superposition of states will always affect the possible outcome of any computation.

Operations

A 3-bit state as well as a 3-qubit state is eight-dimensional vectors. In order to compute in both cases, the system must be initialized into the string corresponding to the vector. If it is a standard randomized computation, the application of stochastic matrices is what helps the system evolve and it preserves that all the probabilities will add up to one, i.e., they preserve L1 norm. When it is quantum computing, only unitary matrices are allowed, i.e., they are effective rotations that preserve the L2 norm where the sum of squares add up to one. The exact unitaries will depend on the quantum device used. Quantum rotations are also reversible since rotations will be undone if rotated backward.

The result has to be read off on termination of an algorithm. In a standard computer a three-bit string is obtained through a sample of the probability distribution in the three-bit register. From a quantum device the three-qubit state would be measured and then sampled from that distribution. The original quantum state is thus destroyed.

For cryptography

A quantum computer would solve a lot of problems that are faced in terms of cryptography in a standard computer. On an ordinary computer, integer factorization for large integers that are a product of a few prime numbers would be infeasible. A quantum computer that uses Shor's algorithm will make this possible. This would mean that a quantum computer could break most of the present-day cryptographic systems that are being used. A polynomial time algorithm would exist to solve the problem. Thus, most public key ciphers based on factoring integers such as the RSA or elliptic curve Diffie-Hellman could be broken using Shor's algorithm. These are popularly used in securing web pages, emails, and other type of data. Quantum computers would have a major effect on electronic security. Algorithms based on other problems rather than discrete logarithm or integer factorization appears to not be broken by such algorithms. Quantum computers have not yet been able to break down lattice-based cryptography or the McEliece cryptosystem. Cryptographic systems based on quantum science would potentially be much more secure against quantum hacking.

For Quantum Database Search

Quantum computers can offer a polynomial speed up for quantum database search. Grover's algorithm can be used for this by utilizing fewer queries than classical algorithms. Grover's algorithm has shown to give the most possible probability of discovering any desired element in oracle searches. This algorithm is useful when there is a collection of many answers but no structure for search. It is also beneficial when the number of inputs in an algorithm is equal to the number of possible answers that have to be checked. Unlike the linear scaling of standard computers, a quantum computer will have a running time that is the square root of the amount of inputs. The application of quantum computing for functions like password cracking is of interest to government bodies in particular.

For Linear Equations

Compared to a standard computer, a quantum computer would be much faster while solving linear equations. This would be more so with the help of the HHL algorithm that is the quantum algorithm used for this purpose. It was named after the scientists Harrow, Hassidim, Lloyd who discovered it.

Obstacles

There are still many obstacles in the path of developing a fully functional quantum-computing device. Building a large-scale device has many technical difficulties as well. IBM plays a major role in this field and has mentioned certain requirements that need to be fulfilled for this. The device needs to be physically scalable in order to increase the quantity of qubits. The qubits need to be initialized to arbitrary values. There has to be a universal gate set as well as easily legible qubits. Quantum gates also need to be faster than decoherence time.

Quantum decoherence can be a major hindrance. It is very challenging to control or remove quantum decoherence. This would require isolation of the system from the environment since interactions with it are what cause the system to decohere. There are also other decoherence sources to consider such as quantum gates and lattice vibrations. Since decoherence is non-unitary it is irreversible. If it cannot be avoided, it should at least be highly controlled. The typical range for decoherence time is nanoseconds to seconds in low temperature. To prevent too much decoherence, some quantum devices require cooling to 20 millikelvins. Due to this, some quantum algorithms will actually be inoperable because they are time-consuming. The superpositions would be corrupted if we try to maintain that state of qubits for too long. The operation usually needs to be completed quicker than decoherence time since error rates are mostly proportional to decoherence rate. According to the quantum threshold theorem, the error rate must be small in order to use quantum error correction to deal with errors or decoherence. If the errors can be corrected faster than decoherence then the total calculation time is usually longer than that of decoherence. In case of depolarizing noise, the required error rate will be $10-3$ for fault tolerant computation.

The scalability condition can be met for many systems. The problem here is that a lot more qubits are required for error correction. For instance, in case of a 1000-bit number, 104 bits will be needed without error correction. The figure would then rise to 107 without error correction. Then the computation time would be 107 steps at 1MHz, 10 seconds. Another approach in the decoherence problem is creating a topological quantum device.

Quantum Computing Models

Many quanta computing models have been developed that are differentiated by the basic elements by which computation will be decomposed. Four models are of practical importance:

- Quantum gate array

- One-way quantum computer

- Adiabatic quantum computer

- Topological quantum computer

Chapter Thirteen: The Biggest Myth about Quantum Physics and other Myths about Quantum Jumps

There are certain rules in our daily lives which we take for granted. How exactly? For instance, cause-and-effect. We have heard about this cause and effect theory about a zillion times by now. Something takes place and it causes a string of events to occur depending upon what exactly happened at the start. It's not just one thing that leads to another. There are several different causes that result in different effects. When we talk about quantum physics, the basic rules are extremely different. It's impossible to define a particular starting point with precision so you can find out the cause. This is because there's a certain kind of uncertainty that is inherently ingrained in our system. And that's why you can't predict anything when it comes to quantum physics. You can't specifically describe how the human system evolves.

At the very least, you can only conclude a set of possibilities that can be calculated. If you are still looking for a definitive observation, interaction, or measurement, then you can experience only a single outcome: the effect you wanted to experience. The very fact that you are making that the effect you were looking for, but the act of making that observation, interaction, or measurement can fundamentally alter the state of your system.

Now, the question is, how does one interpret this behaviour which has been a topic of debate for over 100 years? Or, if it is even possible to interpret it? Regardless of how unsettling the resolution can be for anyone who experiences it. Sounds puzzling? Yes, it is, but it's possible that our interpretations could be the very thing that are preventing us from understanding our quantum reality in depth.

Let's just consider the example of Schrödinger's cat. What really happens when you place a cat in a box along with one single radioactive atom inside it? In the instance that the atom starts decaying, and the poison starts releasing; the cat consumes it and eventually dies. But what if the atom doesn't decay? Then there is no poison released; the cat is saved. Schrodinger was tremendously confused by this particular analogy, because as per popularly believed the cause-and-effect rule; the cat should either be alive or dead. And what caused it? Depending upon whether the atom decayed or not, whether the poison was released or not, or whether the cat lived or died. Think about it for a minute. What if you don't make an observation, measurement or trigger an interaction, which can tell you the exact outcome? Then both the atom (cause) and the cat (effect) must be in a stage of superposition; the cat could be alive or dead, both at the same time. It is this failure to know whether the cat or any other animal for that matter is alive or dead can be considered as a classic example of the weirdness about quantum physics.

For those of you who are still confused, here's another example. Now this isn't exactly an analogy, but an experiment, which is conducted by firing one single electron at a barrier that consists of two separate narrow slits in it. These slits are separated by a short distance and have a screen placed behind them. Now what does your common sense tell you? The first thing that would come to my mind is that the electron must go through either of the two slits - left or right. And when the electrons are in a row, you should be left with two different bunches: the first one which corresponds to the electrons which slipped through right slit, and the second one which corresponds to the electrons which slipped through left slit. Do you know what's surprising? That's not what really happens at all.

Here, what you can see may appear just like an interference pattern. The behaviour of the independent electrons is similar to waves while the pattern appears to be something you would get if the continuous light waves are fired using a double slit. It can also appear as if the water waves are sent through a tank that has two gaps exactly where the slits are.

That said, keep in mind that these are single electrons! At any given point in time, where are they exactly, and which slit did they slip through? Now, some of us may want to make use of a detector at each slit for identifying which slit the electron may slip through. If you chose to use the detector for this purpose, then you can do either one of these things; Number 1 slips through the right slit, then number 2 through the left, number 3 through the left again, number 4 through the right, number 5 through left, and so on. You may think that this will lead you to something conclusive but take a closer look again. When you observe the pattern of electrons on the screen, you won't get the original interference pattern. All you are left with is two different bunches. In short, the case remains non-conclusive? In some weird way, the act of observation, interaction, or measurement has completely changed the outcome. This is what I was talking about when it comes to the unsettling weirdness of quantum physics. There is no clear and single explanation of what is really going on. One of the several approaches used to understand this phenomenon is the creation of an interpretation of quantum mechanics. There have been a lot of ways that people have tried using to make sense of the phenomenon occurring here. Some of them are as follows:

The Copenhagen Interpretation:

This one claims that the quantum wave function is meaningless in physical terms, but that's only until you make a defining measurement. Again, the only thing it does is assign probabilities for what would happen in the event that such a measurement is made that completely "collapses" the wave function.

The Many-Worlds Interpretation:

This one claims that quantum states interact with our environment, which then produces great entanglement along with numerous possible outcomes. This is also where a larger number of different kinds of parallel Universes exist which can help in housing each possible outcome.

The Ensemble Interpretation:

This can be done by imagining numerous identical systems that are prepared in a similar manner. This preparation involves making a measurement by merely choosing one single system and assuming it to be "the real one."

The Pilot Wave/de Broglie-Bohm interpretation:

This one states that the existence and positions of the particles are always guided by wave functions. What does this imply? It means that the wave-guides are not only deterministic, but also appear to be managed by specific hidden variables. These variables could be non-local, which can simultaneously also affect disconnected space-time points.

Despite these various experiments conducted by people across the world, some difficulty arises which is inherent to them: Until now, no one has been able to devise an experiment that can make us discern it from one another. The quantum field theory or the physical theory with regards to quantum mechanics stands all on its own regardless of the type of interpretation you may wish to apply to it.

The fact is that the quantum theory works perfectly well the way it is. This theory has quantum operators acting on the quantum wave functions. This, in-turn, allows a chance at probability distribution of the kind of outcome that might ensue at a later stage. Once you start conducting the relevant experiments, the interpretation you chose to stick to becomes completely irrelevant.

Still, philosophers, physicists, and armchair

students debate about several different interpretations by assuming that they contain different physical meanings. In reality, it may be similar to the age-old story of a bunch of blind men examining an elephant

The realization of Neils Bohr is an interesting one:

He considered the fact that religions over the ages have always been interpreted through parables, images, and paradoxes, and this has made us believe that it's the only way of grasping the reality they intend to offer. Does that mean that it's not a genuine reality? Not at all. And it's not going to help if you decide to split this reality into several objects. Moreover, a subjective interpretation won't get us far either.

I know that each one of us has our favourite interpretations of the quantum theory. How many of our interpretations are really clear and helpful? Most of our interpretations offer nothing but to create confusion instead of throwing light off everything. The various explanations given by different people do not amount to truth. In fact, they only display our limited intuition and human perception. When it comes to making sense out of the quantum universe or fully understanding it, these interpretations do not help us at all. Now, humans can certainly design different types of experiments that can illustrate <u>the behaviour of a specific interpretation</u> of the Universe.

Then, there are questions such as how does quantum physics work? Or does quantum physics work at all? Or what do the mathematical objects included in the theory imply? We can have as many answers and interoperations of them as we care to give them, but all these answers do not constitute one single reality. At best, your answers can say much more about you and your biases, assumption, and prejudices about the Universe rather than the mere reality of the Universe itself. In fact, there aren't more than a few things that we can observe in nature: particle properties such as scattering amplitudes, momentum, position, and cross-sections states are all we have. When you ask questions about the hidden nature of reality it implies that there is one single truth and it assumes that there exists one true reality containing certain rules that fit our intuition. But is this the reality? It turns out that the exact opposite may in fact be true. We might like to think that our perception of reality are the ultimate truth, but we fail to see that we are simply limited by our capabilities and senses. The rules we assume that manage our

Universe could be completely foreign to us than how we choose to conceive it.

Quantum physics is truly fascinating on a lot of levels. The fascination stems from the fact that <u>how differently quantum physics work for us in our daily life experiences.</u> It's surprising how anything and everything can behave as a particle or wave as per what you do it. In short, the Universe is created from indivisible quanta; this means that we, as humans, can only predict the different probabilities of an outcome and not one independent individual outcome. Another fact about quantum physics is that it is completely non-local when it comes to time and space. The effects of which can only be seen and experienced on smaller scales. Inarguably, it's the most puzzling and weirdest thing about the universe that we have discovered.

But even when it's so, we can't resist the temptation of adding ourselves to the equation. This could be because of the allure of the difficult-to-define part of it with regards to interaction, measurement, and observation of it. Humans have always been intrigued by the mystery of the unknown and that explains our fascination with the theory of quantum physics. When we remove ourselves from it, then all we are left with is the results, the equations, and <u>the answers offered by the physical Universe</u>. Physics alone is incapable of answering questions like "why" the Universe functions in a particular manner; at best, it can only offer an idea of how it functions. If you are interested in knowing the basic nature of reality, then ask for answers from the universe and it may soon start spilling out some secrets, and when this happens, listen to it carefully. If you start adding your own stories to it, then chances are that it's your personal interpretations of experiences and are guided by the universe. Under any circumstance, stay away from the temptation of giving it your own meaning which will protect you from falling for the greatest myth about the theory of quantum physics: that it needs to be interpreted at all.

Now that we have debunked the biggest myth about quantum physics, let's look at some of the myths surrounding quantum jumping. Going by the various opinions of different people, here too, there is ample confusion about reality shifts and quantum jumps and what exactly might be in store for anyone who is considering making a quantum jump. Some recent reviews, which were published on <u>Quantum Jumps</u> at Amazon, imply that there is a humongous amount of misunderstanding about what quantum jumps are really all about. Let's have a look at some of the biggest myths with regards to quantum jumps and try to find out if we can bring about any kind of clarification about what's actually going on.

MYTH #1: Quantum jumps only take place in the quantum realm

From where did this myth originate? It stems from the very basic belief of individuals that the term "quantum jump" has actually been misappropriated. These individuals also assume that a quantum jump can only reside in the domain of quantum physics. How true is this? Not true at all. Take for example the English language, where one phrase that appears to have one single meaning later ends up having several different meanings. Just like that certain concepts can be much more centralized compared to others and offers different encounters that are far different from its original meaning. Dr. Warren Nagourney of the University of Washington explains the excitement experienced by a bunch of physicists who have witnessed the quantum jumps themselves.

"You have to hold yourself steady and look for minutes at a time, and then you'll see it switch. You see the trapped ion blinking on and off, and each blink is a quantum jump. It's a striking illustration that things occur discontinuously in nature."

The above observations were made back in 1986. Fast-forward to 2011, there has been some progress in the understanding of the theory of quantum jumps. As per the study published by the physics working with Stanford University's Leonard Susskind and Berkeley University's Raphael Bousso, the findings were as follows:

"... The many worlds of quantum mechanics and the many worlds of the multiverse are the same thing, and that the multiverse is necessary to give exact operational meaning to probabilistic predictions from quantum mechanics."

Myth number 2: The multiverse being un-provable theory is completely non-

scientific

Recently, a particular method was devised with the aim of testing the multiverse by making use of different mathematical models which demonstrate what we could experience if bubble universes collided. In order for a theory to be considered scientific, there needs to be a space that allows you to approve or disapprove a theory. Considering this fact, it's big news that humans will soon be able to identify what they should be looking for astronomically, which matches the results that the computer models show us. It's possible that we will then be in a position to experience what happens when bubble universes crash into each other. Matthew Johnson who works at the Perimeter Institute is on his way to create a <u>computer</u> simulation of our universe. His reports suggest that it's actually quite easy. Fortunately, due to computer simulations, we can safely say that it's possible to rule out specific models of the multiverse. This implies that if at all we are living in a bubble universe, then we might be in a position to confirm the reality of this.

MYTH #3: Quantum jumps facilitates identity theft

There have been some concerns regarding the fact that if there are several versions of you, then it is also possible that several versions of myself exist in this multiverse. If that is the case, then there is no point in messing around with other potential realities and create trouble for yourself. Instead, we should simply keep our focus on our own universe without getting tempted to engage with others. This theory can be extremely helpful if we are able to apply it to our lives. There would be absolutcly nothing that we will want to discover more because we would be creating our own realities; but here's the thing. A lot of people are of the firm opinion that quantum jumps can actually act as a guide for committing cosmic theft. The belief goes something like this, "If you are committing a cosmic theft using quantum jump and causing destruction to another person's life, then it's possible that a different version of you could screw up your life that is better at quantum leaps.

One of the easiest and the best method of clearing this misconception is to refer to the physics papers published by Yasunori Nomura from the UC Berkeley University along with Leonard Susskind and Raphael Bousso. It focuses on a key concept that it could be highly likely that several physicists who have been working on the theories of the world of quantum physics are concluding that it is highly possible everything is just a matter of probabilities. In simple terms, when a person makes a jump to another reality, then he or she is not really messing up anybody's life. In no particular way are they causing others any harm. It's completely possible that it's merely an interpretation of the person on the receiving end based on his or her own experiences.

MYTH #4: One can get trapped in an alternate universe

This is a big one! So many of us have an intense fear of being trapped in an alternate universe without being finding a way to return to our original state. This fear is extremely prominent amongst those who are experiencing significant shifts in their reality. The very concept of a "home" in a multiverse that only exists in probability space suggests that humans are perpetually and physically experiencing a specific universe. This is also similar to expecting many possible outcomes when you travel from one place to another. While you may have different experiences in your own respective universes, you may physically experience only one.

MYTH #5: If you don't experience quick results, then the quantum jumps are not working

This is what messes up when someone is trying to manifest a different experience using quantum jumps. Conquering these abilities requires a certain amount of proficiency, which can only be acquired over time. Momentum is a key thing in being able to reach where you wish to be, but a lot of people get entangled in unrealistic expectations. People who have mastered this art have spent a lot of time and effort learning how to get it right. So, if you are looking for quick results, then quantum jumps might not be your thing. Just as you wouldn't expect to ace a marathon in a week or establish yourself as a world-renowned chef within a month of learning how to cook.

When you first start working on mastering the art of quantum jumping, you may need a lot of patience to be able to hone your internal energy. It can take several weeks, months, or even a year to remove the existing blockages in your body as well as your environment. You will also need to clear yourself of all the fears you have accumulated over the years. This can be done by being present as much as you can. This type of awareness can be brought about by practicing various methods of stillness like meditation, reiki, or many others. Regardless of which method you pick, you need to give this technique enough time for it to start rolling. And even when you start seeing results, you will have to be mindful enough to not constantly measure them. Moving into a constant state of awareness may seem overwhelming at first, but as the time passes by, you will find it easier and easier.

Chapter Thirteen: Summary/ Highlights of Quantum physics

Quantum physics is quite weird. In fact, it happens to be so weird that Neils Bohr said, "Those who are not shocked when they first come across quantum theory cannot possibly have understood it."

By now I am sure you have understood what quantum physics is all about and will not need any special introduction to it. We have read about the different theories relating to it and how it can affect your everyday living. Quantum physics is a vast subject and has many theories and myths attached to it that can be quite intriguing.

Let us now look back at some of the weird aspects associated with quantum physics and brush our memories.

Certain Particles Virtual in nature can appear and disappear randomly.

When we think of empty space, we often assume it is empty, but it is not. Empty space can be full of energy. Random particles can pop in and out of appearance owing to that energy. One aspect of those particles is known as matter and the other element is known as antimatter. Antimatter can be a complex subject with different explanations available. But all that is relevant is that if the particles touch each other then they will explode and then disappear. This will happen about a billionth of a second after they appear and will crash into one another and stop existing. These virtual particles tend to go against a common assumption based on classical theory and common sense that everything that exists must have a cause. Yet, it is not the only assumption with regards to quantum physics.

Superposition- The smaller a thing happens to be, the more likely it is to be present in different places all at once.

Now think of a person standing in a dark room. Once the light is turned on, you will be able to see the person. Until such time, it is mere guesswork as to where exactly the person might be standing in the room. Now let us assume that the person in the room is tiny. If you switch on the lights and switch them off really fast, then they will probably remain in the same spot or maybe not. One switch on-off session could see them in the middle of the room while another could see them in the corner and suddenly, they can be everywhere in the room! Location is nothing but a set of different probabilities that can all exist at the same time. Before quantum particles are measured, they will be forced to pick a location, and set in a state of "superposition" between all the probable locations. This set of locations can probably exist in Broglie wavelength and can get smaller as soon as something gets bigger.

Quantum Tunneling- Micro Particles That Can Teleport

Let's bring back the person in the dark room once again. He/she can be in different locations in the room every time you switch the light off and on again. If walled off from a corner of the room, then they could still appear in the same corner if you switched off the light and on again many times. Although they are usually on one side of the wall, they can also sometimes be on the other side. This might look to you as though the tiny person teleported right through the wall. Just like in the case of superposition, larger things are less likely to teleport compared to smaller ones. If something happens to be randomly jumping from one side to another then it is labeled as quantum tunneling.

Entanglement Is When Tiny Things Can Communicate Faster Than Light

I want you to think of two hills with each one having a person standing on it. The easiest and fastest way for these people to communicate with each other is by using flashlights. This is because light is possibly the fastest thing in the entire universe.

Now let's say the hills and the people on it are really tiny. They shook each other's hands before getting on the hills. This is so that whatever affects one will affect the other. Say one person feels a gush of wind on his face and the other will feel the same! This is assuming that the hills can be super far say billions of miles apart.

Let us now go back to the dark room example. Assume that there are two people in different rooms across the country. One person in one room will be sitting when you switch the light on and the other person across the country will be standing simultaneously.

This is based on the fact that information can travel quite fast and something that Einstein hated and called, "*spooky action at a distance*". He assumed that there could be hidden variables that could explain this without the need to break into the universe's speed limit, but John Bell decided to experiment upon this theory and learned that it was wrong. Entangled particles could communicate at a speed that is faster than light.

There are many different theories about entanglement that do the rounds including that of wormholes where entangled particles are connected physically through tunnels at different points in space-time and the different worlds of quantum physics where every possibility happens with each one set in a different universe. It is yet to be known if these ideas can hold true but, as of now, all we can do is appreciate the ideas that quantum physics has put forth within the last century.

There is possibly no other theory in the physical world that has brought about such thought and research as quantum physics. Not only does it challenge and overthrow previous assumptions of reality but it also showcases extraordinary features that we cannot understand. But we can use it in application to come up with properties of matter with great accuracy and precision that can exceed any current information obtained from experiments.

Conclusion

On that note, I'd like to inform you that we have come to the end of this book. I would like to thank you once again for using this particular source to learn more about quantum physics. I am sure by now your understanding about quantum physics has altered the way you looked at things around you and not make it look like something that only the scientists of the world and the geeks should learn about.

After reading the book, you will now have a much better view of how the field of quantum mechanics started and was developed over decades. This particular field does not have a single scientist to attribute gratitude to but to many who kept questioning and finding new answers to unsolved questions, and there will be many more who will challenge and perhaps change things that we have now accepted based on the previous studies.

Quantum mechanics was the answer for many issues that classical mechanics could not account for. There are still many more questions that must be solved in order for more advancement in this field. The three fundamental principles of quantum theory play a major role in it. You have now taken a peek into the bizarre world of quantum physics. The concepts of quantum physics are usually hard even for scientists to grasp. Even Albert Einstein had many issues with quantum physics since it is not a clear-cut area and shows how random the behavior of matter can be at the subatomic level.

Out of the ones we have mentioned in this text, the important ones that you need to study are the wave-particle duality and the uncertainty principle in particular. Classical physics tends to predict only certainties, but quantum physics has predicted more probabilities. Quantum theory may have many difficult aspects, but it is still a very essential part of modern physics. It can actually be called the most successful theory in science. This particular branch was what paved the way for advancements by application in lasers, electron microscopes, nuclear power, superconductors, transistors, etc. We also learnt more about the physical structure of atoms, properties like chemical bonding, electricity conduction, and the thermal or mechanical properties of solids. We know that quantum theory has only been able to explain some fundamental forces clearly. These are weak nuclear force, electromagnetism, and strong nuclear force.

Physics will move forward over the years when quantum theory is satisfactorily combined with the theory of relativity to give us unified theory. This would be the "theory of everything" that might be able to make sense of everything that exists in the universe.

As the study has progressed over decades, a lot of progress has been made; however, it is still difficult to implement a lot of the possibilities of quantum mechanics in real life. The good news is the work is in progress and will definitely be much more functional in the coming years. You now know a lot more about quantum mechanics, how it was developed over the years and how it can be applied in order to gain much more advancement in various fields. There are comprehensive texts available on the subject that you can easily get access to. This text was a starting point to give you a grasp on the important concepts of quantum mechanics.

A lot of research was done from varied sources to get you all the information you would require in order to better understand the field of quantum physics. By now you have a much clearer picture of the difference between classical physics. The fact that quantum physics deals with the minutiae makes it a very distinct and important part of science.

So, start learning more about it if it interests you and grasp the vast scope of quantum physics. The information gathered here is in very simple language and easy for anyone to comprehend and actually find interesting as well. If you found this text, useful please recommend it to your friends or colleagues interested in quantum physics.

Thank you and good luck!

47869177R00092

Made in the USA
Middletown, DE
11 June 2019